Chaotic Vibrations

Chaotic Vibrations

An Introduction
for Applied Scientists
and Engineers

FRANCIS C. MOON
Theoretical and Applied Mechanics
Cornell University
Ithaca, New York

A WILEY-INTERSCIENCE PUBLICATION
JOHN WILEY & SONS
NEW YORK　CHICHESTER　BRISBANE　TORONTO　SINGAPORE

Library of Congress Cataloging in Publication Data:

Moon, F. C., 1939–
  Chaotic vibrations.

  "A Wiley-Interscience publication."
  Bibliography: p.
  Includes indexes.
  1. Dynamics.   2. Chaotic behavior in systems.
3. Fractals. I. Title.

QA845.M66 1987   531'.11   87-8282
ISBN 0-471-85685-1

Printed in the United States of America

10 9 8 7 6 5 4 3 2 1

To my daughters,
Catherine, Patricia, and Elizabeth,
who have added a little joyful chaos to my life

Preface

Had anyone predicted that new discoveries would be made in dynamics three hundred years after publication of Newton's *Principia*, they would have been thought naive or foolish. Yet in the last decade new phenomena have been observed in all areas of nonlinear dynamics, principal among these being chaotic vibrations. Chaotic oscillations are the emergence of randomlike motions from completely deterministic systems. Such motions had been known in fluid mechanics, but they have recently been observed in low-order mechanical and electrical systems and even in simple one-degree-of-freedom problems. Along with these discoveries has come the recognition that nonlinear difference and differential equations can admit bounded, nonperiodic solutions that behave in a random way even though no random quantities appear in the equations. This has prompted the development of new mathematical ideas, new ways of looking at dynamical solutions, which are now making their way into the laboratory.

It is the purpose of this book to help translate these mathematical ideas and techniques into language that engineers and applied scientists can use to study chaotic vibrations. Although I am an experimenter in dynamics, I have had to acquire a certain level of mathematical understanding of these new ideas, such as strange attractor, Poincaré map, or fractal dimension, in order to study chaotic phenomena in the laboratory. A number of excellent mathematical treatises on chaotic dynamics have appeared recently. I have attempted to read and distill these new concepts with the help of my more theoretical colleagues at Cornell University and attempt in this book to explain the relevance of this new language of dynamics to engineers, especially those who must study and measure vibrations. I believe that these new geometric and topological concepts in dynamics will become part of the laboratory tools in vibration analysis in the same way that Fourier analysis has become a permanent part of engineering experimental technique.

Besides the infusion of new ideas, the study of chaotic vibrations is important to engineering vibrations for several reasons. First, in mechanical systems a chaotic or noisy system makes life prediction or fatigue analysis difficult because the precise stress history in the solid is not known. Second, the recognition that simple nonlinearities can lead to chaotic solutions raises the question of predictability in classical physics and the usefulness of numerical simulation of nonlinear systems. It is part of the conventional wisdom that larger and faster supercomputers will allow one to make more precise predictions of a system's behavior. However, for nonlinear problems with chaotic dynamics, the time history is sensitive to initial conditions and precise knowledge of the future may not be possible even when the motion is periodic.

Many new books on chaotic dynamics assume that the reader has had some exposure to advanced dynamics, nonlinear vibrations, and advanced mathematical techniques. In this book I have tried to work from a background that a B.S. engineering graduate would have; namely, ordinary differential equations, some intermediate-level dynamics, and vibration or systems dynamics courses. I have also tried to give examples of systems with chaotic behavior and to offer engineers the tools to measure, predict, and quantify chaotic vibrations in physical systems.

In Chapter 2 I describe some of the characteristics of chaotic vibrations and how to recognize and test for them in physical experiments. The types of physical models and experimental systems in which chaotic behavior has been observed are given in Chapter 3. In Chapter 4 some experimental techniques are presented to measure chaotic phenomena, including Poincaré maps. This is a "how to do it" chapter and can be skipped by those looking for an overview of the field. Chapters 5 and 6 are more mathematical and explore what criteria now exist to predict when chaotic vibrations will occur and the new concepts in fractal mathematics. Fractal concepts are at the center of many of the new ideas in nonlinear dynamics. Beautiful pictures of fractal geometric objects have appeared in the popular press and have added an aesthetic dimension to the study of dynamics. In Chapter 6 I attempt to relate fractal ideas to specific applied problems in nonlinear dynamics.

One might ask: Why write this book now while the field of nonlinear vibrations is undergoing such rapid change? First, it was an opportune time since I was asked to prepare and deliver eight lectures on chaotic vibrations at the Institute for Fundamental Technical Problems in Warsaw, Poland, in August 1984. This book is an outgrowth of those lectures. Second, during

1984–1985 I was invited to give lectures in chaotic vibrations at nearly thirty universities and research laboratories. Many colleagues expressed a desire for a book on chaos, aimed at those in the applied sciences. Also, I felt that many engineers in the field of vibrations were unaware of the exciting new things happening in dynamics. Engineering researchers, armed with new tools of dynamical systems, will I am sure make further advances into this new area by exploring new applications and developing more practical tools for measuring and describing these new phenomena.

I want to thank my colleagues at Cornell University in theoretical and applied mechanics, especially Philip Holmes and Richard Rand, who have patiently tried to explain these new mathematical ideas to me. I have also had useful conversations with John Guckenheimer, formerly of the University of California at Santa Cruz but now at Cornell. The deliberate lack of rigor in this book in describing some of the new geometric and topological concepts must be blamed on me, however. I have proceeded on the assumption that the book will succeed only if it stimulates interest in this new field. Given this stimulation, I hope the reader will seek out more mathematical references to provide more detailed and precise discussion of these new ideas.

Finally I wish to recognize the contributions of graduate students and research associates who have worked so enthusiastically with me on problems of chaos: Joseph Cusumano, Mohammed Golnaraghi, Guang-Xian Li, Chih-Kung Lee, Bimal Poddar, Gabriel Raggio, and Stephan Shaw (now at Michigan State University). Special mention is also made of the technical help of Stephen King and William Holmes who helped design some of the electronic instrumentation in our experiments on chaotic vibration.

Regarding the references at the end of this book, I did not attempt to include all the historically significant papers in chaotic studies and I apologize to those researchers whose fine contributions to the subject have not been cited. The inclusion of more of my own papers than those of others must be interpreted as an author's vanity and not any measure of their relative importance to the field.

I also want to acknowledge funding from the National Science Foundation through the solid mechanics program under Dr. Clifford Astill, from the Air Force Office of Scientific Research through Dr. Anthony Amos of the Aerospace Section, from the Office of Naval Research through Dr. Michael Shlesinger of the Physics Division, from the Army Research Office through Dr. Gary Anderson of the Engineering Sciences Division and from the IBM Corporation.

FRANCIS C. MOON

Ithaca, New York
May 1987

Contents

Chaotic Vibrations

1

Introduction: A New Age of Dynamics

In the beginning, how the heav'ns and earth rose out of chaos
J. Milton, *Paradise Lost*, 1665

1.1 WHAT IS CHAOTIC DYNAMICS?

For some, the study of dynamics began and ended with Newton's Law of $F = mA$. We were told that if the forces between particles and their initial positions and velocities were given, one could predict the motion or history of a system forever into the future, given a big enough computer. However, the arrival of large and fast computers has not fulfilled the promise of infinite predictability in dynamics. In fact, it has been discovered quite recently that the motion of very simple dynamical systems cannot always be predicted far into the future. Such motions have been labeled *chaotic* and their study has prompted a discussion of some exciting new mathematical ideas in dynamics. With the approaching tricentennial of Newton's *Principia* (1687), in which he introduced the calculus into the study of dynamics, it is appropriate that three centuries later new phenomena have been discovered in dynamics and that new mathematical concepts from topology and geometry have entered this venerable science.

1

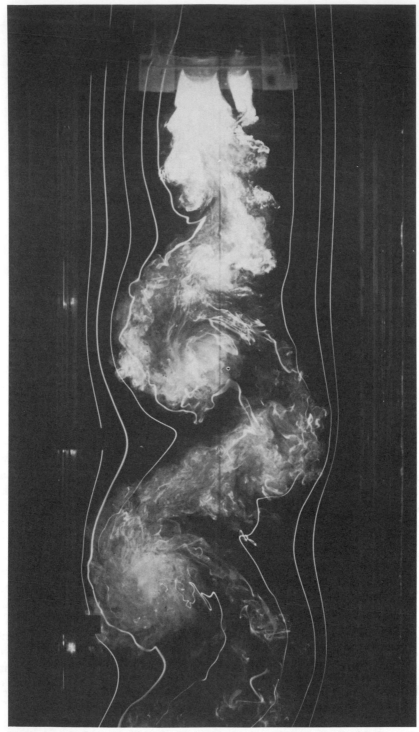

Figure 1-1 Turbulent wake in the flow past a circular cylinder [courtesy of R. Dumas].

The nonscientific concept of chaos[1] is very old and often associated with a physical state or human behavior without pattern and out of control. The term chaos often stirs fear in humankind since it implies that governing laws or traditions no longer have control over events such as prison riots, civil war, or a world war. Yet there is always the hope that some underlying force or reason is behind the chaos or can explain why seemingly random events appear unpredictable.

In the physical sciences, the paragon of chaotic phenomena is turbulence. Thus, a rising column of smoke or the eddies behind a boat or aircraft wing[2] provide graphic examples of chaotic motion (Figure 1-1). The fluid mechanician, however, believes that these events are not random because the governing equations of physics for each fluid element can be written down. Also, at low velocities, the fluid patterns are quite regular and predictable from these equations. Beyond a critical velocity, however, the flow becomes turbulent. A great deal of the excitement in nonlinear dynamics today is centered around the hope that this transition from ordered to disordered flow may be explained or modeled with relatively simple mathematical equations. What we hope to show in this book is that these new ideas about turbulence extend to solid mechanical and electrical continua as well. It is the recognition that chaotic dynamics are inherent in all of nonlinear physical phenomena that has created a sense of revolution in physics today.

We must distinguish here between so-called random and chaotic motions. The former is reserved for problems in which we truly do not know the input forces or we only know some statistical measures of the parameters. The term chaotic is reserved for those deterministic problems for which there are no random or unpredictable inputs or parameters. The existence of chaotic or unpredictable motions from the classical equations of physics was known by Poincaré.[3] Consider the following excerpt from

---

[1] The origin of the word *chaos* is a Greek verb which means *to gape open* and which was often used to refer to the primeval emptiness of the universe before things came into being (*Encyclopaedia Britannica*, Vol. 5, p. 276). To the stoics, chaos was identified with water and the watery state which follows the periodic destruction of the earth by fire. Ovid in *Metamorphises* used the term to denote the raw and formless mass in which all is disorder and from which the ordered universe is created. A modern dictionary definition of chaos (Funk and Wagnalls) provides two meanings: (i) utter disorder and confusion and (ii) the unformed original state of the universe.

[2] The reader should look at the beautiful collection of photos of fluid turbulent phenomena compiled by Van Dyke (1982).

[3] Henri Poincaré (1854–1912) was a French mathematician, physicist, and philosopher whose career spanned the grand age of classical mechanics and the revolutionary ideas of relativity and quantum mechanics. His work on problems of celestial mechanics led him to questions of dynamic stability and the problem of finding precise mathematical formulas for the dynamic

this essay on *Science and Method*:

> It may happen that small differences in the initial conditions produce very great ones in the final phenomena. A small error in the former will produce an enormous error in the latter. Prediction becomes impossible.

In the current literature, *chaotic* is a term assigned to that class of motions in deterministic physical and mathematical systems whose time history has a *sensitive dependence on initial conditions*.

Two examples of mechanical systems that exhibit chaotic dynamics are shown in Figure 1-2. The first is à thought experiment of an idealized billiard ball (rigid body rotation is neglected) which bounces off the sides of an elliptical billiard table. When elastic impact is assumed, the energy remains conserved, but the ball may wander around the table without exactly repeating a previous motion for certain elliptically shaped tables.

Another example, which the reader with access to a laboratory can see for oneself, is the ball in a two-well potential shown in Figure 1-2*b*. Here the ball has two equilibrium states when the table or base does not vibrate. However, when the table vibrates with periodic motion of large enough amplitude, the ball will jump from one well to the other in an apparently random manner; that is, periodic input of one frequency leads to a randomlike output with a broad spectrum of frequencies. The generation of a continuous spectrum of frequencies below the single input frequency is one of the characteristics of chaotic vibrations (Figure 1-3).

Loss of information about initial conditions is another property of a chaotic system. Suppose one has the ability to measure a position with accuracy Δx and a velocity with accuracy Δv. Then in the position–velocity plane (known as the phase plane) we can divide up the space into areas of size $\Delta x \, \Delta v$ as shown in Figure 1-4. If we are given initial conditions to the stated accuracy, we know the system is somewhere in the shaded box in the phase plane. But if the system is chaotic, this uncertainty grows in time to $N(t)$ boxes as shown in Figure 1-4*b*. The growth in uncertainty given by

$$N \approx N_0 e^{ht} \qquad (1\text{-}1.1)$$

is another property of chaotic systems. The constant h is related to the concept of *entropy* in information theory (e.g., see Shaw, 1981, 1984) and will also be related to another concept called the *Lyapunov exponent* (see

---

history of a complex system. In the course of this research he invented the "the method of sections," now known as the Poincaré section or map.

An excellent discussion of uncertainties and determinism and Poincaré's ideas on these subjects may be found in the very readable book by L. Brillouin (1964, Chapter IX).

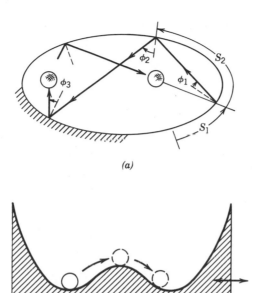

(a)

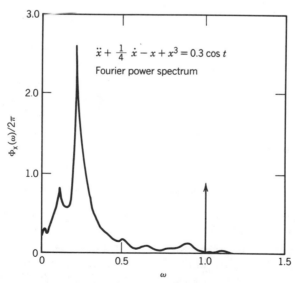

(b)

Figure 1-2 (*a*) The motion of a ball after several impacts with an elliptically shaped billiard table. The motion can be described by a set of discrete numbers (s_i, ϕ_i) called a map. (*b*) The motion of a particle in a two-well potential under periodic excitation. Under certain conditions, the particle jumps back and forth in a periodic way, that is, LRLR $\cdots$, or LLRLLR $\cdots$, and so on, and for other conditions the jumping is chaotic that is, it shows no pattern in the sequence of symbols L and R.

$$\ddot{x} + \tfrac{1}{4}\,\dot{x} - x + x^3 = 0.3 \cos t$$

Fourier power spectrum

Figure 1-3 The power spectral density (Fourier transform) of chaotic motion in a two-well potential (after Y. Udea, Kyoto University).

5

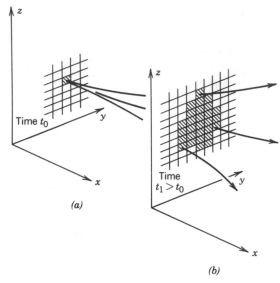

Figure 1-4 An illustration of the growth of uncertainty or loss of information in a dynamical system. The black box at time $t = t_0$ represents the uncertainty in initial conditions.

Chapter 5) which measures the rate at which nearby trajectories of a system in phase space diverge.

The reader may ask: With predictability lost in chaotic systems, is there any order left in the system? For dissipative systems the answer is yes; there is an underlying structure to chaotic dynamics. This structure is not apparent by looking at the dynamics in the conventional way, that is, the output versus time or from frequency spectra. One must search for this order in phase space (position versus velocity). There one will find that chaotic motions exhibit a new geometric property called *fractal* structure. One of the goals of this book is to teach how to discover the fractal structure in chaotic vibrations as well as to measure the loss of information in these randomlike motions.

Why Should Engineers Study Chaotic Dynamics?

Recently, the subject of chaos has become newsworthy—the study of mathematical chaos that is. Many popular science magazines and even *The New York Times* and *Newsweek* have carried articles on the new studies into mathematics of chaotic dynamics. But engineers have always known about chaos—it was called noise or turbulence and fudge factors or factors of safety were used to design around these apparent random unknowns that seem to crop up in every technical device. So what is new about chaos?

First, the recognition that chaotic vibrations can arise in low-order, nonlinear deterministic systems raises the hope of understanding the source of randomlike noise and doing something about it. Second, the new discoveries in nonlinear dynamics bring with them new concepts and tools for detecting chaotic vibrations in physical systems and for quantifying this "deterministic noise" with new measures such as fractal dimensions and Lyapunov exponents.

Mathematicans have also known that certain dynamical systems possessed irregular solutions since around the turn of the century. Poincaré, as noted in the above quote, was aware of chaotic solutions, as was Birkhoff in the early part of this century. Van der Pol and Van der Mark (1927) reported "irregular noise" in experiments with an electronic oscillator in the magazine *Nature*. So what is new about chaos?

What is new about chaotic dynamics is the discovery of a seemingly underlying order which holds out the promise of being able to predict certain properties of noisy behavior. Perhaps the greatest hope lies in the possibility of understanding turbulence in fluid, thermofluid, and thermochemical systems. Turbulence is one of the few remaining unsolved problems of classical physics, and the recent discovery of deterministic systems which exhibit chaotic oscillations has created much optimism about solving the mysteries of turbulence. But already this optimism has been tempered by the complexities of chaotic dynamics in thermofluid systems. However, there may be more immediate payoffs in the study of chaotic phenomena in systems with fewer degrees of freedom, such as low-order nonlinear mechanical devices and nonlinear circuits.

Sources of Chaos in Continuum Physics

Chaotic vibrations occur when some strong nonlinearity exists. Examples of nonlinearities in mechanical and electrical systems include the following:

Nonlinear elastic or spring elements
Nonlinear damping such as friction
Backlash, play, or limiters or bilinear springs
Fluid-related forces
Nonlinear boundary conditions
Nonlinear feedback control forces in servosystems
Nonlinear resistive, inductive, or capacitive circuit elements
Diodes
Many transistors and other active devices
Electric and magnetic forces

In mechanical continua, nonlinear effects arise from a number of different sources which include the following:

1. Kinematics; for example, convective acceleration, Coriolis and centripetal accelerations
2. Constitutive relations, for example, stress versus strain
3. Boundary conditions, for example, free surfaces in fluids, deformation-dependent constraints
4. nonlinear body forces, for example, magnetic or electric forces
5. Geometric nonlinearities associated with large deformations in structural solids such as beams plates and shells

How such nonlinearities enter the laws of mechanics can be seen by looking at the equation of momentum balance in continuum mechanics,

$$\nabla \cdot \mathbf{t} + \mathbf{f} = \frac{\partial \mathbf{v}}{\partial t} + \mathbf{v} \cdot \nabla \mathbf{v} \tag{1-1.2}$$

where $\mathbf{t}$ is the stress tensor, $\mathbf{f}$ is the body force, and the right-hand side represents the acceleration. Nonlinearities can enter this equation through the stress–strain or stress–strain rate relations in the first left-hand term. Nonlinear body forces such as occur in magnetohydrodynamics or plasma physics can enter the body force term $\mathbf{f}$. Finally, on the right-hand side of Eq. (1-1.2), we see an explicit nonlinear term in the convective acceleration. This term appears in many fluid flow problems and is one of the sources of turbulence in fluids.

While chaotic phenomena have been observed in thermofluid, solid mechanical, and electrical systems, chaotic fluid phenomena have sometimes been portrayed as *fundamental* because of the ubiquitous and pervasive nature of turbulence. However, in the classic Navier–Stokes equations of fluid mechanics, derived from the momentum balance equation (1-1.2), one can see that the nonlinearity resides in the convective acceleration or kinematic term:

$$\nu \nabla^2 \mathbf{v} - \nabla P = \frac{\partial \mathbf{v}}{\partial t} + \mathbf{v} \cdot \nabla \mathbf{v} \tag{1-1.3}$$

where ν is the kinematic viscosity and P is the pressure. The viscous term on the left-hand side is linear and based on the assumption of a Newtonian fluid.

One can imagine that, if one goes beyond the study of the Navier–Stokes equation to include nonlinear viscous fluids (non-Newtonian fluids) or elastoplastic materials, there is probably a vast array of nonlinear and chaotic phenomena to be discovered in mechanics, electromagnetics, and acoustics. Thus, there should be no claim that convective acceleration terms represent some fundamental nonlinearity in classical physics.

Where Have Chaotic Vibrations Been Observed?

From the previous discussion, one can see that chaotic phenomena can be observed in many physical systems. Each month new phenomena are reported in the scientific and engineering literature. A partial list of mechanical and electrical systems known to exhibit chaotic vibrations includes the following:

Vibrations of buckled elastic structures

Mechanical systems with play or backlash

Aeroelastic problems

Wheel–rail dynamics in rail systems

Magnetomechanical actuators

Large, three-dimensional vibrations of structures such as beams and shells

Systems with sliding friction

Rotation or gyroscopic systems

Nonlinear acoustic systems

Simple forced circuits with diodes

Harmonically forced circuits with p-n transistor elements

Harmonically forced circuits with nonlinear capacitance and inductance elements

Feedback control devices

Laser and nonlinear optical systems

These are but a few of the many phenomena in which chaos has been uncovered. Descriptions of specific examples are given in Chapter 3. A question asked by most novices to the field of chaotic dynamics is: If chaos is so pervasive, why was it not seen earlier in experiments? Two responses to this question come to mind. First, if one goes back and reads earlier papers on experiments in nonlinear vibrations, one often finds a brief mention of nonperiodic phenomena buried in a discussion of more classical nonlinear vibrations (see Chapter 3 for examples). Second, Joseph Keller, an applied mathematician at Stanford University, in responding to this question in a lecture speculates that earlier scientists and engineers were taught almost exclusively in linear mathematical ideas, including linear algebra and differential equations. Hence, it was natural, Keller summarizes, that when approaching dynamic experiments in the laboratory, they looked only for phenomena that fit the linear mathematical models.

As to why theorists had not come upon those ideas earlier, there is evidence that some did, like Poincaré and Birkhoff. However, specific manifestations of chaotic solutions had to wait for the arrival of powerful computers with which to calculate the long time histories necessary to observe and measure chaotic behavior.

1.2 CLASSICAL NONLINEAR VIBRATION THEORY: A BRIEF REVIEW

In this section, we present a short review of classical vibration theory, linear and nonlinear. This is meant simply to define and review a few ideas in nonlinear dynamics concerning periodic vibration so we may later be able to contrast these with chaotic vibration. Readers desiring more detailed discussion in classical nonlinear vibration should consult books such as Stoker (1950), Minorsky (1962), or Nayfeh and Mook (1979). We begin with a brief review of linear vibration concepts.

Linear Vibration Theory

The classic paradigm of linear vibrations is the spring–mass system shown in Figure 1-5 along with its electric circuit analog. When there is no disturbing force, the undamped system vibrates with a frequency that is independent of the amplitude of vibration:

$$\omega_0 = \left(\frac{k}{m} \right)^{1/2} = \left(\frac{1}{LC} \right)^{1/2} \qquad (1\text{-}2.1)$$

In this state, energy flows alternatively between elastic energy in the spring (electric energy in the capacitor C) and kinetic energy in the mass (magnetic energy in the inductor L). The addition of damping ($c \neq 0$, $R \neq 0$) introduces decay in the free vibrations so that the amplitude of the mass (or charge in the circuit) exhibits the following time dependence:

$$x(t) = A_0 e^{-\gamma t}\cos\left[\left(\omega_0^2 - \gamma^2\right)^{1/2}t + \varphi_0\right] \qquad (1\text{-}2.2)$$

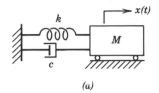

(a)

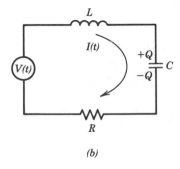

R

(b)

Figure 1-5 (*a*) The classic, mechanical spring–mass–dashpot oscillator. (*b*) The electrical circuit analog of a damped oscillator.

where

$$\gamma = \frac{c}{2m} \quad \text{or} \quad \gamma = \frac{R}{2L}$$

The system is said to be *underdamped* when $\gamma^2 < \omega_0^2$, *critically damped* when $\gamma^2 = \omega_0^2$, and *overdamped* when $\gamma^2 > \omega_0^2$.

One of the classic phenomena of linear vibratory systems is that of *resonance* under harmonic excitation. For this problem, the differential equation that models the system is of the form (e.g., see Thompson, 1965)

$$\ddot{x} + 2\gamma\dot{x} + \omega_0^2 x = f_0 \cos \Omega t \tag{1-2.3}$$

If one fixes f_0 and varies the driving frequency Ω, the absolute magnitude of the steady-state displacement of the mass (after transients have damped out) reaches a maximum close to the natural frequency ω_0, or more precisely at $\Omega = (\omega_0^2 - \gamma^2)^{1/2}$. This phenomenon is sketched in Figure 1-6. The effect is more pronounced when the damping is small. This is indeed the case in structural systems and engineers are familiar with the problem of fatigue failures in structures and machines owing to large resonance-excited vibrations. When a linear mechanical system has many degrees of freedom, one often models it as a coupled set of spring–mass oscillators leading to the phenomena of multiple resonant frequencies when the system is harmonically forced. This behavior has often led vibration analysts to

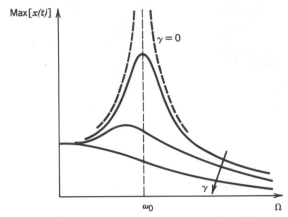

Figure 1-6 Classical resonance curves (response amplitude versus frequency) for the forced motion of a damped *linear* oscillator for different values of the damping γ.

assume that every peak in a vibration frequency spectrum is associated with at least one mode or degree of freedom. In nonlinear vibrations, this is not the case. A one-degree-of-freedom nonlinear system can generate many frequency spectra in contrast to its linear counterpart, as was shown in Figure 1-3. In any event, the mathematical theory of linear systems is well understood and has been codified in sophisticated computer software packages. Nonlinear problems are another story.

Nonlinear Vibration Theory

Nonlinear effects can enter the problem in many ways. A classical example is a nonlinear spring where the restoring force is not linearly proportional to the displacement. For the case of a symmetric nonlinearity (equal effects for compression or tension), the equation of motion takes the following form:

$$\ddot{x} + 2\gamma\dot{x} + \alpha x + \beta x^3 = f(t) \qquad (1\text{-}2.4)$$

When the system is undamped and $f(t) = 0$, periodic solutions exist where the natural frequency increases with amplitude for $\beta > 0$. This model is often called a *Duffing* equation, after the mathematican who studied it.

If the system is acted on by a periodic force, in the classical theory one assumes that the output will also be periodic. When the output has the same frequency as the force, the resonance phenomenon for the nonlinear spring is shown in Figure 1-7. If the amplitude of the forcing function is held

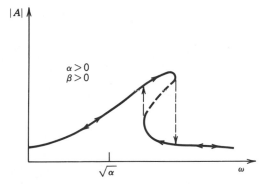

Figure 1-7 Classical resonance curve for a *nonlinear* oscillator with a hard spring when the response is periodic with the same period as the driving force. [α and β refer to Eq. (1-2.4).]

constant, there exists a range of forcing frequencies for which three possible output amplitudes are possible as shown in Figure 1-7. One can show that the dashed curve in Figure 1-7 is unstable so that a *hysteretic effect* occurs for increasing and decreasing frequencies. This is called a *jump phenomenon* and can be observed experimentally in many mechanical and electrical systems.

Other periodic solutions can also be found such as *subharmonic* and *superharmonic* vibrations. If the driving force has the form $f_0 \cos \omega t$, then a subharmonic oscillation may take the form $x_0 \cos(\omega t / n + \varphi)$ plus higher harmonics (n is an integer). Subharmonics play an important role in prechaotic vibrations as we shall see later.

Nonlinear resonance theory depends on the assumption that periodic input yields periodic output. However, it is this postulate that has been challenged in the new theory of chaotic vibrations.

Self-excited oscillations are another important class of nonlinear phenomena. These are oscillatory motions which occur in systems that have no periodic inputs or periodic forces. Several examples are shown in Figure 1-8. In the first, the friction created by relative motion between a mass and moving belt leads to oscillations. In the second example there exists the whole array of aeroelastic vibrations in which the steady flow of fluid past a solid object on elastic restraints produces steady-state oscillations. A classic electrical example is the vacuum tube circuit studied by Van der Pol and shown in Figure 1-9.

In each case, there is a steady source of energy, a source of dissipation, or a nonlinear restraining mechanism. In the case of the Van der Pol oscillator, the source of energy is a dc voltage. It manifests itself in the

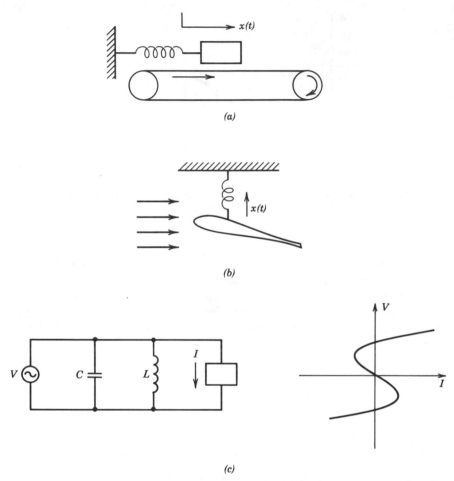

Figure 1-8 Example of self-excited oscillations: (*a*) dry friction between a mass and moving belt; (*b*) aeroelastic forces on a vibrating airfoil; and (*c*) negative resistance in an active circuit element.

mathematical model of the circuit as a negative damping:

$$\ddot{x} - \gamma\dot{x}\left(1 - \beta x^2\right) + \omega_0^2 x = 0 \qquad (1\text{-}2.5)$$

For low amplitudes, energy can flow into the system, but at higher amplitudes the nonlinear damping limits the amplitude.

In the case of the Froude pendulum (e.g., see Minorsky, 1962, Chap. 28), the constant rotation of the shaft provides an energy input. For small

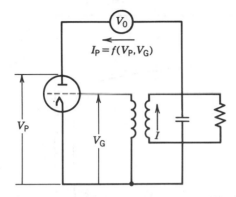

Figure 1-9 Sketch of a vacuum tube circuit with limit cycle oscillation of the type studied by Van der Pol.

vibrations, the nonlinear friction is modeled as negative damping; while for large vibrations, the amplitude of the vibration is limited by the nonlinear term $\beta\dot{\theta}^3$:

$$\ddot{\theta} + \alpha \sin \theta = T_0 + \gamma\dot{\theta}\left(1 - \beta\dot{\theta}^2\right) \tag{1-2.6}$$

The oscillatory motions of such systems are often called *limit cycles*. The phase plane trajectories for the Van der Pol equation are shown in Figure 1-10. Small motions spiral out to the closed asymptotic trajectory, while large motions spiral onto the limit cycle. (In Figures 1-10, 1-11, $y = \dot{x}$.)

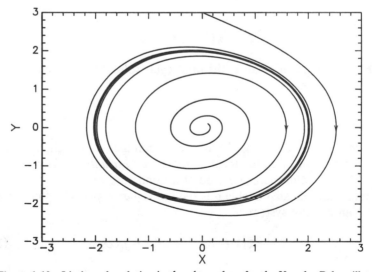

Figure 1-10 Limit cycle solution in the phase plane for the Van der Pol oscillator.

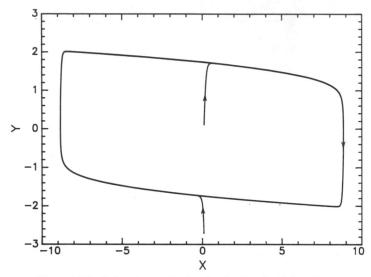

Figure 1-11   Relaxation oscillation for the Van der Pol oscillator.

Two questions are often asked when studying problems of this kind: (1) What is the amplitude and frequency of the limit cycle vibrations? (2) For what parameters will stable limit cycles exist?

In the case of the Van der Pol equation, it is convenient to normalize the dependent variable by $\sqrt{\beta}$ and the time by ω_0^{-1} so the equation assumes the form

$$\ddot{x} - \epsilon\dot{x}(1 - x^2) + x = 0 \qquad (1\text{-}2.7)$$

where $\epsilon = \gamma/\omega_0$. For small ϵ, the limit cycle solution is a circle of radius 2 in the phase plane; that is,

$$x \simeq 2\cos t + \cdots \qquad \text{`}(1\text{-}2.8)$$

where the $+ \cdots$ indicates third-order harmonics and higher. When ϵ is larger the motion takes the form of *relaxation oscillations* shown in Figure 1-11 with a nondimensional period of around 1.61 when $\epsilon > 10$.

A more complicated problem is the case when a periodic force is added to the Van der Pol system:

$$\ddot{x} - \gamma\dot{x}(1 - \beta x^2) + \omega_0^2 x = f_0\cos\omega_1 t \qquad (1\text{-}2.9)$$

Since the system is nonlinear, *superposition of free and forced oscillations is*

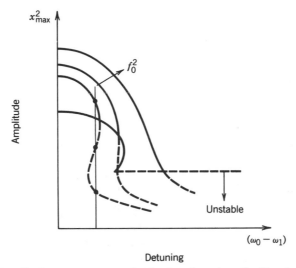

Figure 1-12 Amplitude response curves for the forced motion of a Van der Pol oscillator (1-2.9).

not valid. Instead, if the driving frequency is close to the limit cycle frequency, the resulting periodic motion will become *entrained* at the driving frequency. For small forcing, there are three possible periodic solutions, but only one is stable as shown in Figure 1-12. For larger forces f_0, there is only one solution. In either case, when f_0 is fixed, the entrained periodic solution becomes unstable when the detuning parameter $\omega_0 - \omega_1$ is increased and other motions become possible.

When the difference between the driving and free oscillation frequencies is large, a new phenomenon is possible in the Van der Pol system—*combination oscillations*—sometimes called almost periodic or quasiperiodic solutions. Combination oscillation solutions take the form

$$x = b_1 \cos \omega_1 t + b_2 \cos \omega_2 t \qquad (1\text{-}2.10)$$

When ω_1 and ω_2 are incommensurate, that is, ω_1/ω_2 is an irrational number, the solution is said to be *quasiperiodic*. In the case of the Van der Pol equation (1-2.9), $\omega_2 \equiv \omega_0$ which is the free oscillation limit cycle frequency (e.g., see Stoker, 1950, p. 166).

More will be said about quasiperiodic vibrations later, but since they are not periodic, they may be mistaken for chaotic solutions, which they are not. [For one, the Fourier spectrum of Eq. (1-2.10) is just two spikes at $\omega = \omega_1, \omega_2$, whereas for chaotic solutions the spectrum is broad and continuous.]

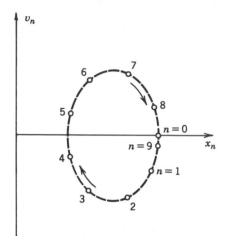

Figure 1-13 Stroboscopic plot of quasiperiodic solutions of the Van der Pol equation (1-2.9) in the phase plane.

The phase plane portrait of Eq. (1-2.10) is not closed when ω_1 and ω_2 are incommensurate, so another method is used to portray the quasiperiodic function graphically. To do this, we stroboscopically sample $x(t)$ with a period equal to $2\pi/\omega_1$; that is, let

$$t_n = \frac{n2\pi}{\omega_1} \tag{1-2.11}$$

and denote $x(t_n) = x_n$, $\dot{x}(t_n) = v_n$. Then Eq. (1-2.10) becomes

$$x_n = b_i + b_1\cos\frac{2\pi n\omega_0}{\omega_1}, \qquad v_n = -\omega_0 b_2\sin\frac{2\pi n\omega_0}{\omega_1} \tag{1-2.12}$$

As n increases, the points (x_n, v_n) move around an ellipse in the stroboscopic phase plane (called a Poincaré map), as shown in Figure 1-13. When ω_0/ω_1 is incommensurate, the set of points $\{x_n, v_n\}$ for $n \to \infty$ fill in a closed curve given by

$$\left(x_n - b_i\right)^2 + \left(\frac{v_n}{\omega_0}\right)^2 = b_2^2 \tag{1-2.13}$$

Quasiperiodic oscillations also occur in systems with more than one degree of freedom.

Local Geometric Theory of Dynamics

Modern ideas about nonlinear dynamics are often presented in geometric terms or pictures. For example, the motion of an undamped oscillator, $\ddot{x} + \omega_0 x = 0$, can be represented in the phase plane $(x, \dot{x})$ by an ellipse (Figure 1-13). In this picture, time is implicit and the time history runs clockwise around the ellipse. The size of the ellipse depends on the given initial conditions for $(x, \dot{x})$.

More generally for nonlinear problems, one first finds the equilibrium points of the system and examines the motion around each equilibrium point. The local motion is characterized by the nature of the eigenvalues of the linearized system. Thus, if the dynamical model can be represented by a set of first-order differential equations

$$\dot{\mathbf{x}} = \mathbf{f}(\mathbf{x}) \qquad (1\text{-}2.14)$$

where x represents a vector whose components are the state variables, then the *equilibrium points* are defined by $\dot{\mathbf{x}} = 0$, or

$$\mathbf{f}(\mathbf{x}_e) = 0 \qquad (1\text{-}2.15)$$

For example, in the case of the harmonic oscillator, there is just one equilibrium point at the origin $\mathbf{x} = (x, v \equiv \dot{x})$, $x_e = 0$, $v_e = 0$. To determine the nature of the dynamics about $\mathbf{x} = \mathbf{x}_e$, one expands the function $\mathbf{f}(\mathbf{x})$ in a Taylor series about each equilibrium point $\mathbf{x}_e$ and examines the dynamics of the linearized problem.

To illustrate the method, consider the set of two first-order equations:

$$\begin{aligned} \dot{x} &= f(x, y) \\ \dot{y} &= g(x, y) \end{aligned} \qquad (1\text{-}2.16)$$

When time does not appear explicitly in the functions $f(\ )$ and $g(\ )$, the problem is called *autonomous*. The equilibrium points must satisfy two equations: $f(x_e, y_e) = 0$ and $g(x_e, y_e) = 0$. Introducing small variables about each equilibrium point, that is,

$$x = x_e + \eta \quad \text{and} \quad y = y_e + \xi$$

the linearized system can be written in the form

$$\frac{d}{dt}\begin{Bmatrix} \eta \\ \xi \end{Bmatrix} = \begin{bmatrix} \dfrac{\partial f}{\partial x} & \dfrac{\partial f}{\partial y} \\[2mm] \dfrac{\partial g}{\partial x} & \dfrac{\partial g}{\partial y} \end{bmatrix} \begin{Bmatrix} \eta \\ \xi \end{Bmatrix} \qquad (1\text{-}2.17)$$

where the derivatives are evaluated at the point (x_e, y_e).

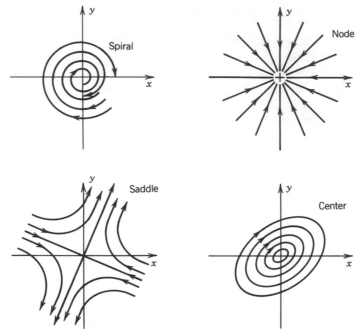

Figure 1-14 Classical phase plane portraits near four different types of equilibrium points for a system of two time-independent differential equations.

Some authors use the notation $\nabla \mathbf{F}$ or $D\mathbf{F}$, where $\mathbf{F} = (f, g)$, to represent the matrix of partial derivatives in Eq. (1-2.17). The nature of the motion about each equilibrium point is determined by looking for eigensolutions

$$\begin{Bmatrix} \eta \\ \xi \end{Bmatrix} = \begin{Bmatrix} \alpha \\ \beta \end{Bmatrix} e^{st} \tag{1-2.18}$$

where α and β are constants. The motion is classified according to the nature of the two eigenvalues of $D\mathbf{F}$ [i.e., whether s is real or complex and whether Real $(s) > 0$ or < 0.]

Sketches of trajectories in the phase plane for different eigenvalues are shown in Figure 1-14. For example, the *saddle point* is obtained when both eigenvalues s are real, but $s_1 < 0$ and $s_2 > 0$. A *spiral* occurs when s_1 and s_2 are complex conjugates.

The *stability* of the linearized system (1-2.17) depends on the sign of Real (s). When one of the real parts of s_1 and s_2 is positive, the motion about the equilibrium point is *unstable*. If the roots are not pure imaginary numbers, then theorems exist to show that the local motion of the linearized system is qualitatively similar to the original nonlinear system (1-2.16). Pure oscillatory motion in the linearized system ($s = \pm i\omega$) requires further analysis to establish the stability of the nonlinear system. These ideas for a second-order system can be generalized to higher-dimensional phase spaces (e.g., see Arnold, 1978 or Guckenheimer and Holmes, 1983).

Bifurcations

As parameters are changed in a dynamical system, the stability of the equilibrium points can change as well as the number of equilibrium points. The study of these changes in nonlinear problems as system parameters are varied is the subject of *bifurcation theory*. Values of these parameters at which the qualitative or topological nature of motion changes are known as critical or bifurcation values.

As an example, consider the solutions to the undamped Duffing oscillator

$$\ddot{x} + \alpha x + \beta x^3 = 0 \qquad (1\text{-}2.19)$$

One can first plot the equilibrium points as a function of α. As α changes from positive to negative, one equilibrium point splits into three points. Dynamically, one center is transformed into a saddle point at the origin and two centers (Figure 1-15). This kind of bifurcation is known as a *pitchfork*. Physically, the force $-(\alpha x + \beta x^3)$ can be derived from a potential energy function. When α becomes negative, a one-well potential changes into a double-well potential problem. This represents a qualitative change in the dynamics and thus $\alpha = 0$ is a critical bifurcation value.

Another example of a bifurcation is the emergence of limit cycles in physical systems. In this case, as some control parameter is varied, a pair of complex conjugate eigenvalues $s_1, s_2 = \pm i\omega + \gamma$ cross from the left-hand plane ($\gamma < 0$, a stable spiral) into the right-hand plane ($\gamma > 0$, an unstable spiral) and a periodic motion emerges known as a *limit cycle*. This type of qualitative change in the dynamics of a system is known as a *Hopf bifurcation* and is illustrated in Figure 1-16.

The theory we have just described is called a *local* analysis because it only tells what happens dynamically in the vicinity of each equilibrium point. The pièce de resistance in classical dynamical analysis is to piece together all the local pictures and describe a *global* picture of how trajectories move between and among equilibrium points.

Such analysis is tractable when bundles of different trajectories corresponding to different mutual conditions move more or less together as a laminar fluid flow. Such is the case when the phase space has only *two*

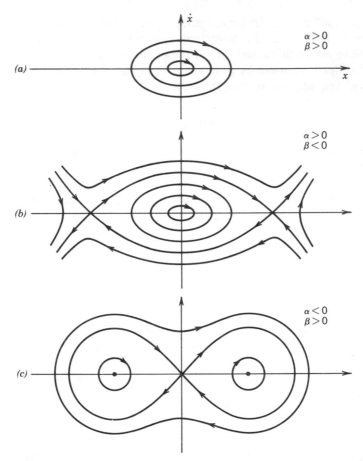

Figure 1-15 Phase plane trajectories for an oscillator with a nonlinear restoring force [Duffing's equation (1-2.19)]: (*a*) Hard spring problem; $\alpha, \beta > 0$. (*b*) Soft spring problem; $\alpha > 0$, $\beta < 0$. (*c*) Two-well potential; $\alpha < 0$, $\beta > 0$.

dimensions. However, when there are three or more first-order equations, the bundles of trajectories can split apart and get tangled up into what we now call chaotic motions.

From this brief review, one can see that there are three classic types of dynamical motion:

1. Equilibrium
2. Periodic motion or a limit cycle
3. Quasiperiodic motion

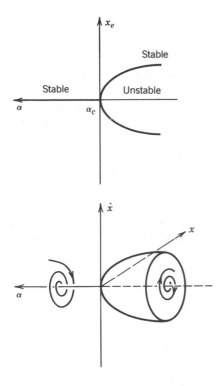

Figure 1-16 Bifurcation diagrams: (*a*) Pitchfork bifurcation for Duffing's equation (1-2.19)—transition from one to two stable equilibrium positions. (*b*) Hopf bifurcation—transition from stable spiral to limit cycle oscillation.

These states are called *attractors*, since if some form of damping is present the transients decay and the system is "attracted" to one of the above three states. The purpose of this book is to describe another class of motions in nonlinear vibrations that is not one of the above classic attractors. This new class of motions is chaotic, in the sense of not being predictable when there is a small uncertainty in the initial condition, and is often associated with a state of motion called a *strange attractor*.

The classic attractors are all associated with classic geometric objects in phase space, the equilibrium state with a point, the periodic motion or limit cycle with a closed curve, and the quasiperiodic motion with a surface in a three-dimensional phase space. The "strange attractor," as we shall see in later chapters, is associated with a new geometric object (new relative to what is now taught in classical geometry) called a *fractal set*. In a three-dimensional phase space, the fractal set of a strange attractor looks like a collection of an infinite set of sheets or parallel surfaces, some of which are separated by distances that approach the infinitesimal. This new attractor in nonlinear dynamics requires new mathematical concepts and a language to describe it as well as new experimental tools to record it and give it some quantitative measure. The relationship between bifurcations and chaos is discussed in a recent book by Thompson and Stewart (1986).

1.3 MAPS AND FLOWS

Mathematical models in dynamics generally take one of three forms: differential equations (or *flows*), difference equations (called *maps*), and *symbol dynamic equations*.

The term flow refers to a bundle of trajectories in phase space originating from many contiguous initial conditions. The continuous time history of a particle is the most familiar example of a flow to those in engineering vibrations. However, certain qualitative and quantitative information may be obtained about a system by studying the evolution of state variables at discrete times. In particular, in this book we shall discuss how to obtain difference evolution equations from continuous time systems through the use of the Poincaré section. These Poincaré maps can sometimes be used to distinguish between various qualitative states of motion such as periodic, quasiperiodic, or chaotic. In some problems not only time is restricted to discrete values, but knowledge of the state variables may be restricted to a finite set of values or categories such as red or blue or zero or one. For example, in the double-well potential of Figure 1-2*b*, one may be interested only in whether the particle is in the left or right well. Thus, an orbit in time may consist of a sequence of symbols LRRLRLLLR $\cdots$. A periodic orbit might be LRLR $\cdots$ or LLRLLR $\cdots$. In the new era of nonlinear dynamics, all three types of model are used to describe the evolution of physical systems. (See Crutchfield and Packard (1982) or Wolfram (1986), for a discussion of symbol dynamics.)

In a periodically forced vibratory system, a Poincaré map may be obtained by stroboscopically measuring the dynamic variables at some particular phase of the forcing motion. In an *n*-state variable problem, one can obtain a Poincaré section by measuring the $n - 1$ variables when the *n*th variable reaches some particular value or when the phase space trajectory crosses some arbitrary plane in phase space as shown in Figure 1-17 (see also Chapters 2 and 4). If one has knowledge of the time history between two penetrations of this plane, one can relate the position at t_{n+1} to that at t_n through given functions. For example, for the case shown in Figure 1-17,

$$\xi_{n+1} = f(\xi_n, \eta_n) \quad \text{and} \quad \eta_{n+1} = g(\xi_n, \eta_n) \qquad (1\text{-}3.1)$$

The mathematical study of such maps is similar to that for differential equations. One can find equilibrium or fixed points of the map and one can classify these fixed points by the study of linearized maps about the fixed point. If $\mathbf{x}_{n+1} = \mathbf{f}(\mathbf{x}_n)$ is a general map of say n variables represented by

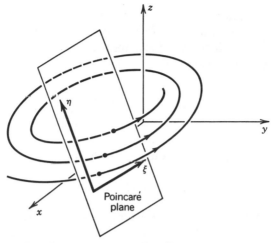

Figure 1-17 Poincaré section: construction of a difference equation model (map) from a continuous time dynamical model.

the vector **x**, then a fixed point satisfies

$$\mathbf{x}_e = \mathbf{f}(\mathbf{x}_e) \tag{1-3.2}$$

The iteration of a map is often written $\mathbf{f}(\mathbf{f}(\mathbf{x})) = \mathbf{f}^{(2)}(\mathbf{x})$. Using this notation, an "m-cycle" or m-periodic orbit is a fixed point that repeats after m iterations of the map; that is,

$$\mathbf{x}_0 = \mathbf{f}^{(m)}(\mathbf{x}_0) \tag{1-3.3}$$

Implied in these ideas is the notion that periodic motions in continuous time history show up as fixed points in the difference equations obtained from the Poincaré sections. Thus, the most generally accepted paradigms for the study of the transition from periodic to chaotic motions is the study of simple one-dimensional and two-dimensional maps.

Three Paradigms for Chaos

Perhaps the simplest example of a dynamic model that exhibits chaotic dynamics is the logistic equation or population growth model (e.g., see May, 1976):

$$x_{n+1} = ax_n - bx_n^2 \tag{1-3.4}$$

The first term on the right-hand side represents a growth or birth effect, while the nonlinear term accounts for the limits to growth such as availability of energy or food. If the nonlinear term is neglected ($b = 0$), the linear equation has an explicit solution:

$$x_{n+1} = ax_n; \qquad x_n = x_0 a^n \qquad (1\text{-}3.5)$$

This solution is stable for $|a| < 1$ and unstable for $|a| > 1$. In the latter case, the linear model predicts unbounded growth which is unrealistic.

The nonlinear model (1-3.4) is usually cast in a nondimensional form:

$$x_{n+1} = \lambda x_n (1 - x_n) \qquad (1\text{-}3.6)$$

This equation has equilibrium points at $x = 0, 1$. For $\lambda > 1$, two equi-

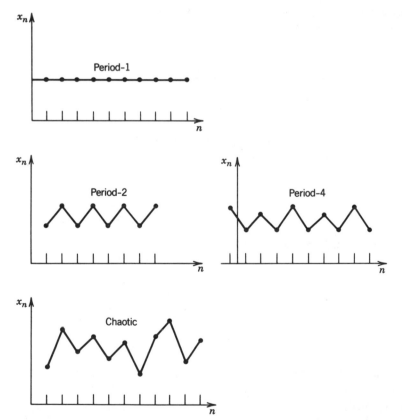

Figure 1-18   Possible solutions to the quadratic map [logistic equation (1-3.6)]. *Top*: Steady period-1 motion. *Middle*: Period-2 and period-4 motions. *Bottom*: Chaotic motions.

librium points exist [i.e., $x = \lambda x(1 - x)$]. To determine the stability of a map $x_{n+1} = f(x_n)$, one looks at the value of the slope $|f'(x)|$ evaluated at the fixed point. The fixed point is unstable if $|f'| > 1$. In the case of the logistic equation (1-3.6) when $1 < \lambda < 3$, there are two fixed points $x = 0$, $(\lambda - 1)/\lambda$ of which the origin is unstable and the other point is stable.

For $\lambda = 3$, however, the slope at $x = (\lambda - 1)/\lambda$ becomes greater than one $(f' = 2 - \lambda)$ and both equilibrium points become unstable. For parameter values of λ between 3 and 4, this simple difference equation exhibits many multiple-period and chaotic motions. At $\lambda = 3$, the steady solution becomes unstable, but a two-cycle or double-period orbit becomes stable. This orbit is shown in Figure 1-18. The value of x_n repeats every two iterations.

For further increases of λ, the period-two orbit becomes unstable and a period-four cycle emerges, only to bifurcate to a period-eight cycle for a higher value of λ. This period-doubling process continues until λ approaches the value $\lambda_\infty = 3.56994 \cdots$. Near this value, the sequence of period-doubling parameter values scales according to a precise law:

$$\frac{\lambda_{n+1} - \lambda_n}{\lambda_n - \lambda_{n-1}} \rightarrow 4.66920 \cdots \tag{1-3.7}$$

The limit ratio is called the Feigenbaum number, named after the physicist who discovered the properties of this map.

Beyond λ_∞ chaotic iterations can occur; that is, the long-term behavior does not settle down to any simple periodic motion. There are also certain narrow windows $\Delta\lambda$ for $\lambda_\infty < \lambda < 4$ for which periodic orbits exist. Periodic and chaotic orbits of the logistic map are shown in Figure 1-19 by plotting x_{n+1} versus x_n.

This map is not only useful as a paradigm for chaos, but it has been shown that other maps $x_{n+1} = f(x_n)$, in which $f(x)$ is double or multiple valued, behave in a similar manner with the same scaling law (1-3.7). Thus, the phenomenon of period doubling or bifurcation parameter scaling has been called a *universal* property for certain classes of one-dimensional difference equation models of dynamical processes.

Period doubling and Feigenbaum scaling (1-3.7) have been observed in many physical experiments (see Chapter 3). This suggests that for many continuous time history processes, the reduction to a difference equation through the use of the Poincaré section has the properties of the quadratic map (1-3.4): hence the importance of maps to the study of differential equations. [See Chapter 5 for a further discussion of the logistic equation (1-3.4).]

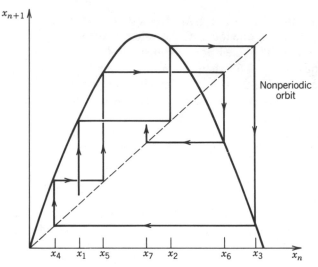

Figure 1-19 Graphical solution to a first-order difference equation. The example shown is the quadratic map (1-3.6).

Henon and Horseshoe Maps

Of course, most physical systems require more than one state variable, and it is necessary to study higher-dimensional maps. One extension of the Feigenbaum problem (1-3.6) is a two-dimensional map proposed by Henon (1976), a French astronomer:

$$x_{n+1} = 1 - \alpha x_n^2 + y_n$$

$$y_{n+1} = \beta x_n$$

(1-3.8)

Note that if $\beta = 0$, we recover the quadratic map. When $|\beta| < 1$, the map contracts areas in the xy plane. It also stretches and bends areas in the phase plane as illustrated in Figure 1-20. This stretching, contraction, and bending or folding of areas in phase space is analogous to the making of a horseshoe. Multiple iterations of such *horseshoe maps* lead to complex orbits in phase space and loss of information about initial conditions and chaotic behavior.

An illustration of the ability of a simple map to produce complex motions is provided in Figure 1-21. In one iteration of the map, a rectangular area is stretched in the vertical direction, contracted in the horizontal

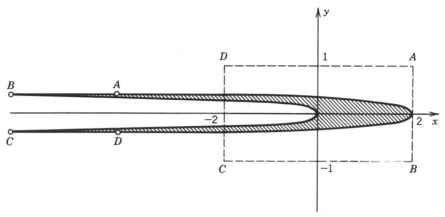

Figure 1-20 Transformation of a rectangular collection of initial conditions under an iteration of the second-order set of difference equations called a Henon map (1-3.8) showing stretching, contraction, and folding which leads to chaotic behavior ($\alpha = 1.4$, $\beta = 0.3$).

direction, and folded or bent into a horseshoe and placed over the original area. Thus, points originally in the area get mapped back onto the area, except for some points near the bend in the horseshoe. If one follows a group of nearby points after many iterations of this map, the original neighboring cluster of points gets dispersed to all sectors of the rectangular area. This is tantamount to a loss of information as to where a point originally started from. Also, the original area gets mapped into a finer and finer set of points, as shown in Figure 1-21. This structure has a fractal property that is a characteristic of a chaotic attractor which has been labeled "strange." This fractal property of a strange attractor is illustrated in the Henon map, Figure 1-22. Blowups of small regions of the Henon attractor reveal finer and finer structure. This self-similar structure of chaotic attractors can often be revealed by taking Poincaré maps of experimental chaotic oscillators (see Chapter 4). The fractal property of self-similarity can be measured using a concept of fractal dimension, which is discussed in Chapter 6.

It is believed by some mathematicians that horseshoe maps are fundamental to most chaotic differential and difference equation models of dynamic systems (e.g., see Guckenheimer and Holmes, 1983). This idea is the centerpiece of a method developed to find a criterion for when chaotic vibrations are possible in a dynamical system and when predictability of future time history becomes sensitive to initial conditions. This Melnikov method has been used successfully to develop criteria for chaos for certain problems in one-degree-of-freedom nonlinear oscillation (e.g., see Chapter 5).

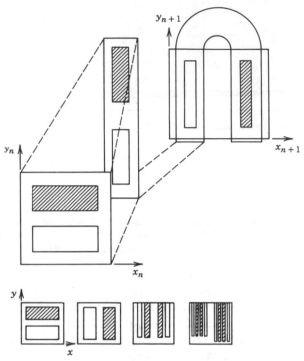

Figure 1-21 The horseshoe map showing how stretching, contraction, and folding leads to fractal-like properties after many iterations of the map.

The Lorenz Attractor and Fluid Chaos

For many readers, the preceding discussion on maps and chaos may not be convincing as regards unpredictability in real physical systems. And were it not for the following example from fluid mechanics, the connection between maps, chaos, and differential equations of physical systems might still be buried in mathematics journals. In 1963, an atmospheric scientist named E. N. Lorenz of M.I.T. proposed a simple model for thermally induced fluid convection in the atmosphere.[4] Fluid heated from below becomes lighter and rises, while heavier fluid falls under gravity. Such motions often produce convection rolls similar to the motion of fluid in a circular torus as shown in Figure 1-23. In Lorenz's mathematical model of convection, three

[4]Lorenz credits Saltzman (1962) with actually discovering nonperiodic solutions to the convection problem in which he used a system of five first-order equations. Mathematicians, however, chose instead to study Lorenz's simpler third-order set of equations (1-3.9). Thus flows the course of scientific destinies.

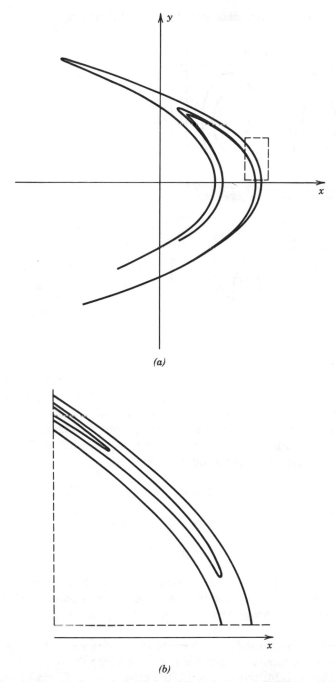

Figure 1-22 (*a*) The locus of points for a chaotic trajectory of the Henon map ($\alpha = 1.4$, $\beta = 0.3$). (*b*) Enlargement of strange attractor showing finer fractal-like structure.

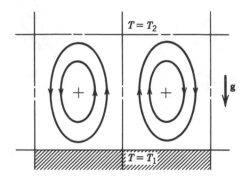

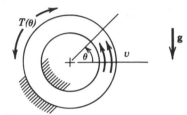

Figure 1-23 *Top*: Sketch of fluid streamlines in a convection cell for steady motions. *Bottom*: One-dimensional convection in a circular tube under gravity and thermal gradients.

state variables are used (x, y, z). The variable x is proportional to the amplitude of the fluid velocity circulation in the fluid ring, while y and z measure the distribution of temperature around the ring. The so-called Lorenz equations may be derived formally from the Navier–Stokes partial differential equations of fluid mechanics (e.g., see Chapter 3). The nondimensional forms of Lorenz's equations are

$$\dot{x} = \sigma(y - x)$$

$$\dot{y} = \rho x - y - xz \qquad (1\text{-}3.9)$$

$$\dot{z} = xy - \beta z$$

The parameters σ and ρ are related to the Prandtl number and Rayleigh number, respectively, and the third parameter β is a geometric factor. Note that the only nonlinear terms are xz and xy in the second and third equations.

For $\sigma = 10$ and $\beta = 8/3$ (a favorite set of parameters for experts in the field), there are three equilibria for $r > 1$ for which the origin is an unstable saddle (Figure 1-24). When $r > 25$, the other two equilibria become unstable spirals and a complex chaotic trajectory moves between all three equilibria as shown in Figure 1-25. It was Lorenz's insistence in the years

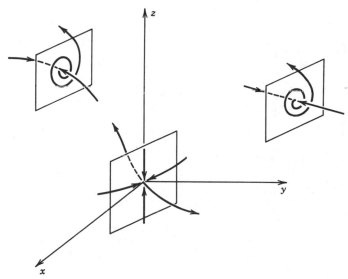

Figure 1-24 Sketch of local motion near the three equilibria for the Lorenz equations (1-3.9).

following 1963 that such motions were not artifacts of computer simulation but were inherent in the equations themselves that led mathematicians to study these equations further (e.g., see Sparrow, 1982). Since 1963, hundreds of papers have been written about these equations and this example has become a classic model for chaotic dynamics.

Systems of other third-order equations have since been found to exhibit chaotic behavior. For example, the forced motion of a nonlinear oscillator can be written in a form similar to (1-3.9); Newton's law for a particle under a force $F(x, t)$ is written

$$m\ddot{x} = F(x, t) \tag{1-3.10}$$

To put (1-3.10) into a form for phase space study, we write $y = \dot{x}$. Furthermore, if the mass is periodically forced, one can reduce the second-order nonautonomous system (1-3.10) to an autonomous system of third-order equations. Thus, we assume

$$F(x, t) = m(f(x, y) + g(t))$$

$$g(t + \tau) = g(t)$$

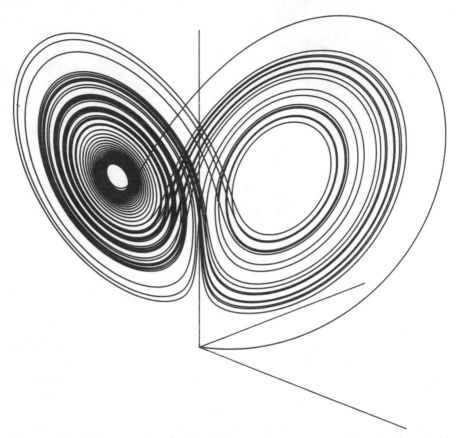

Figure 1-25 Trajectory of a chaotic solution to the Lorenz equations for thermofluid convection (1.3-9) (numerical integration).

By defining $z = \omega t$ and $\omega = 2\pi/\tau$, the resulting equations become

$$\dot{x} = y$$

$$\dot{y} = f(x, y) + g(z) \qquad (1\text{-}3.11)$$

$$\dot{z} = \omega$$

A specific case that has strong chaotic behavior is the Duffing oscillator $F = -(ax + bx^3 + cy)$ (see Chapters 2 and 3).

It is worth noting that, for a two-dimensional phase space, solutions to autonomous systems cannot exhibit chaos since the solution curves of the "flow" cannot cross one another. However, in the forced oscillator or the three-dimensional phase space, these curves can become "tangled" and chaotic motions are possible.

Closing Comments

Dynamics is the oldest branch of physics. Yet 300 years after publication of Newton's *Principia* (1687), new discoveries are still emerging. The ideas of Euler, Lagrange, Hamilton, and Poincaré that followed, once conceived in the context of planetary mechanics, have now transcended all areas of physics. As the new science of dynamics gave birth to the calculus in the 17th century, so today modern nonlinear dynamics has ushered in new

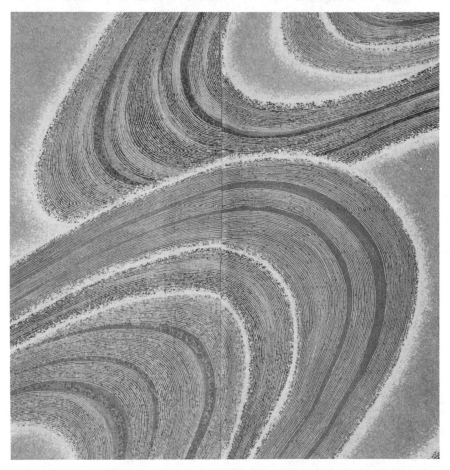

Figure 1-26   Fractal-like pattern in a Japanese kimono design (courtesy of Mitsubishi Motor Corp.).

ideas of geometry and topology, such as fractals, which the 20th century scientist must master to grasp the subject fully.

The ideas of chaos go back in Western thought to the Greeks. But these ideas centered on the order in the world that emerged from a formless chaotic, fluid world in prehistory. G. Mayer-Kress (1985) of Los Alamos National Laboratory has pointed out that the idea of chaos in Eastern thought, such as Taoism, was associated with patterns within patterns, eddies within eddies as occur in the flow of fluids (e.g., see the Japanese kimono design in Figure 1-26).

The view that order emerged from an underlying formless chaos and that this order is recognized only by predictable periodic patterns was the predominant view of 20th century dynamics until the last decade. What is replacing this view is the concept of chaotic events resulting from orderly laws, not a formless chaos, but one in which there are underlying patterns, fractal structures, governed by a new mathematical view of our "orderly" world.

2

How to Identify
Chaotic Vibrations

By the terms of the law of periodic repetition nothing
whatsoever can happen a single time only, everything
happens again, and yet again, and still
again—monotonously. Nature has no originality...
Mark Twain, *Letters from the Earth*

In this chapter, we present a set of diagnostic tests that can help to identify chaotic oscillations in physical systems. Although this chapter is written primarily for those not trained in the mathematical theory of dynamics, theoreticians may find it of interest to see how theoretical ideas about chaos are realized in the laboratory. In a later chapter (Chapter 5), we present some predictive criteria as well as more sophisticated diagnostic tests for chaos. These, however, require some mathematical background, such as the theory of fractal sets (Chapter 6) and Lyapunov exponents (Chapter 5).

Engineers often have to diagnose the source of unwanted oscillations in physical systems. The ability to classify the nature of oscillations can provide a clue as to how to control them. For example, if the system is thought to be *linear*, large periodic oscillators may be traced to a *resonance* effect. However, if the system is *nonlinear*, a *limit cycle* may be the source of periodic vibration, which in turn may be traced to some *dynamic instability* in the system.

To identify nonperiodic or chaotic motions, the following checklist is provided:

(a) Identify nonlinear element in the system.
(b) Check for sources of random input in the system.

(c) Observe time history of measured signal.

(d) Look at phase plane history.

(e) Examine Fourier spectrum of signal.

(f) Take Poincaré map of signal.

(g) Vary system parameters (routes to chaos).

In later chapters, we discuss more advanced techniques. These include measuring two properties of the motion: (1) fractal dimension and (2) Lyapunov exponents.

In the following, we go through the above cited checklist and describe the characteristics of chaotic vibrations. To focus the discussion, the vibration of the buckled beam (double-well potential problem) is used as an example to illustrate the characteristics of chaotic dynamics.

A diagnosis of chaotic vibrations implies that one has a clear definition of such motions. However, as research uncovers more complexities in nonlinear dynamics, a rigorous definition seems to be limited to certain classes of mathematical problems. For the experimentalist, this presents a difficulty since his or her goal is to discover what mathematical model best fits the data. Thus, at this stage of the subject, we use a collection of diagnostic criteria as well as a variety of classes of chaotic motions (see Table 2-1). The experimentalist is encouraged to use two or more tests to obtain a consistent picture of the chaos.

To help sort out the growing definitions and classes of chaotic motions, we list the most common attributes without mathematical formulas, but with the most successful diagnostic tools in parentheses.

CHARACTERISTICS OF CHAOTIC VIBRATIONS

Sensitivity to changes in initial conditions (often measured by Lyapunov exponent, [Chapter 5], and fractal basin boundaries, [Chapter 6])

Broad spectrum of Fourier transform when motion is generated by a single frequency (measured by fast Fourier transform—or FFT—using modern electronic spectrum analyzers)

Fractal properties of the motion in phase space which denote a strange attractor (measured by Poincaré maps, fractal dimensions [Chapter 6])

Increasing complexity of regular motions as some experimental parameter is changed for example, period doubling (often the Feigenbaum number can be measured [Chapters 1 and 5])

Transient or intermittent chaotic motions; nonperiodic bursts of irregular motion (intermittency) or initially randomlike motion that eventually settles down into a regular motion (measurement techniques are few but include the average lifetime of the chaotic burst or transient

TABLE 2-1   Classes of Motion in Nonlinear Deterministic Systems

Regular Motion—Predictable: Periodic oscillations, quasiperiodic motion; not sensitive to changes in parameters or initial conditions

Regular Motion—Unpredictable: Multiple regular attractors (e.g., more than one periodic motion possible); long-time motion sensitive to initial conditions

Transient Chaos: Motions that look chaotic and appear to have characteristics of a strange attractor (as evidenced by Poincaré maps) but that eventually settle into a regular motion

Intermittent Chaos: Periods of regular motion with transient bursts of chaotic motion; duration of regular motion interval unpredictable

Limited or Narrow-Band Chaos: Chaotic motions whose phase space orbits remain close to some periodic or regular motion orbit; spectra often show narrow or limited broadening of certain frequency spikes

Large-Scale or Broad-Band Chaos—Weak: Dynamics can be described by orbits in a low-dimensional phase space $3 \leq n < 7$ (1–3 modes in mechanical systems) and usually one can measure fractal dimensions < 7; chaotic orbits traverse a broad region of phase space; spectra show broad range of frequencies especially below the driving frequency (if one is present)

Large-Scale Chaos—Strong: Dynamics must be described in a high-dimensional phase space; large number of essential degrees of freedom present; difficult to measure reliable fractal dimension; dynamical theories currently unavailable

as some parameter is varied. The scaling behavior might suggest the correct mathematical model; see Chapter 5.)

(a)   Nonlinear System Elements

A chaotic system must have nonlinear elements or properties. A *linear system cannot exhibit chaotic vibrations*. In a linear system, periodic inputs produce periodic outputs of the same period after the transients have decayed (Figure 2-1). (Parametric linear systems are an exception.) In mechanical systems, nonlinear effects include the following:

1   Nonlinear elastic or spring elements
2. Nonlinear damping, such as stick–slip friction
3. Backlash, play, or bilinear springs
4. Most systems with fluids
5. Nonlinear boundary conditions

Nonlinear elastic effects can reside in either material properties or geometric effects. For example, the relation between stress and strain in

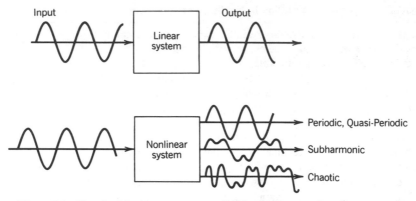

Figure 2-1   Sketch of the input–output possibilities for linear and nonlinear systems.

rubber is nonlinear. However, while the stress–strain law for steel is usually linear below yield, large displacement bending of a beam, plate, or shell may exhibit nonlinear relations between the applied forces or moments and displacements. Such effects in mechanics due to large displacements or rotations are usually called geometric nonlinearities.

In electromagnetic systems, nonlinear properties arise from the following:

1. Nonlinear resistive, inductive, or capacitive elements
2. Hysteretic properties of ferromagnetic materials
3. Nonlinear active elements such as vacuum tubes, transistors, and diodes
4. Moving media problems: for example, $\mathbf{v} \times \mathbf{B}$ voltages, where $\mathbf{v}$ is a velocity and $\mathbf{B}$ is the magnetic field
5. Electromagnetic forces: for example, $\mathbf{F} = \mathbf{J} \times \mathbf{B}$, where $\mathbf{J}$ is current and $\mathbf{F} = \mathbf{M} \cdot \nabla \mathbf{B}$, where $\mathbf{M}$ is the magnetic dipole strength

Common electric circuit elements such as diodes and transistors are examples of nonlinear devices. Magnetic materials such as iron, nickel, or ferrites exhibit nonlinear constitutive relations between the magnetizing field and the magnetic flux density. Some investigators have created negative resistors with bilinear current–voltage relations by using operational amplifiers and diodes (see Chapter 4).

The task of identifying nonlinearities in the system may not be easy: First, because we are often trained to think in terms of linear systems; and second, the major components of the system could be linear but the

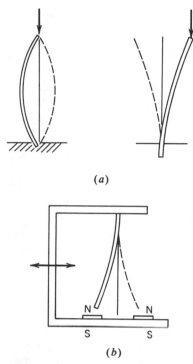

(a)

(b)

Figure 2-2 Nonlinear, multiple equilibrium state problems: (a) buckling of a thin elastic beam-column due to axial end loads and (b) buckling of an elastic beam due to nonlinear magnetic body forces.

nonlinearity arises in a subtle way. For example, the individual elements of a truss structure could be linearly elastic, but the way in which they are fastened together could have play and nonlinear friction present; that is, the nonlinearities could reside in the boundary conditions.

In the example of the buckled beam, identification of the nonlinear element is easy (Figure 2-2). Any mechanical device that has more than one static equilibrium position either has play, backlash, or nonlinear stiffness. In the case of the beam buckled by end loads (Figure 2-2a), the geometric nonlinear stiffness is the culprit. If the beam is buckled by magnetic forces (Figure 2-2b), the nonlinear magnetic forces are the sources of chaos in the system.

(b) Random Inputs

In classical, linear, random vibration theory, one usually treats a model of a system with random variations in the applied forces or model parameters of the form

$$[m_0 + m_1(t)]\ddot{x} + [c_0 + c_1(t)]\dot{x} + [k_0 + k_1(t)]x = f_0(t) + f_1(t)$$

where $m_1(t)$, $c_1(t)$, $k_1(t)$, and $f_1(t)$ are assumed to be random time functions with given statistical measures such as the mean or standard deviation. One then attempts to calculate the statistical properties of $x(t)$ in terms of the given statistical measures of the random inputs. In chaotic vibrations, there are no *assumed* random inputs; that is, the applied forces or excitations are assumed to be deterministic.

By definition, chaotic vibrations arise from deterministic physical systems or deterministic differential or difference equations. While noise is always present in experiments, even in numerical simulations, it is presumed that large nonperiodic signals do not arise from very small input noise. Thus, a *large output signal to input noise ratio is required* if one is to attribute nonperiodic response to a deterministic system behavior.

(c) Observation of Time History

Usually, the first clue that the experiment has chaotic vibrations is the observation of the signal amplitude with time on a chart recorder or oscilloscope (Figure 2-3). The motion is observed to exhibit no visible pattern or periodicity. This test is *not foolproof*, however, since a motion could have a long-period behavior that is not easily detected. Also, some nonlinear systems exhibit quasiperiodic vibrations where two or more incommensurate periodic signals are present. Here the signal may appear to be nonperiodic but it can be broken down into the sum of two or more periodic signals.

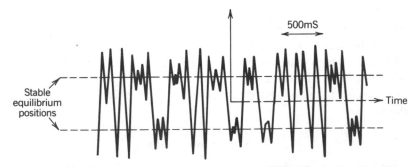

Figure 2-3 Time history of chaotic motions of a buckled elastic beam showing jumps between the two stable equilibrium states.

(d) Phase Plane History

Consider a one-degree-of-freedom mass with displacement $x(t)$ and velocity $v(t)$. Its equations of motion, derived from Newton's law, can be written in the form

$$\dot{x} = v$$

$$\dot{v} = \frac{1}{m} f(x, v, t) \tag{2-1}$$

where m is the mass and f is the applied force. The phase plane is defined as the set of points (x, v). (Some authors use the momentum mv instead of v.) When the motion is periodic (Figure 2-4a), the phase plane orbit traces out a closed curve which is best observed on an analog oscilloscope. For example, the forced oscillations of a linear spring–mass–dashpot system exhibit an elliptically shaped orbit. However, a forced nonlinear system with a cubic spring element may show an orbit that crosses itself but is still closed. This can represent a subharmonic oscillation.

Systems for which the force does not depend explicitly on time, for example, $f = f(x, v)$ in Eq. (2-1), are called *autonomous*. For autonomous nonlinear systems (no harmonic inputs), periodic motions are referred to as *limit cycles* and also show up as closed orbits in the phase plane (see Chapter 1).

Chaotic motions, on the other hand, have orbits that never close or repeat. Thus, the trajectory of the orbits in the phase plane will tend to fill up a section of the phase space as in Figure 2-4b. Although this wandering of orbits is a clue to chaos, continuous phase plane plots provide very little information and one must use a modified phase plane technique called Poincaré maps (see below).

Often, one has only a single measured variable $v(t)$. If $v(t)$ is a velocity variable, one can integrate to get $x(t)$ so that the phase plane consists of points $[\int_0^t v \, d\tau, v(t)]$. On the other hand, if one has to differentiate a displacement- or strain-related signal $x(t)$, high-frequency noise is often introduced. In this case, the experimenter is advised to use a good low-pass filter in $x(t)$ before differentiation (see Chapter 4).

Pseudo-Phase-Space Method. Another technique that has been used when only one variable is measured is the time-delayed pseudo-phase-plane method (also called the embedding space method). For a one degree-of-freedom system with measurement $x(t)$, one plots the signal versus itself but delayed or advanced by a fixed time constant: $[x(t), x(t + T)]$. The idea is that the signal $x(t + T)$ is related to $\dot{x}(t)$ and should have

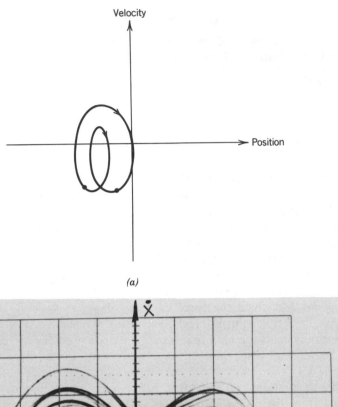

Velocity

Position

(a)

$\dot{X}$

X

(b)

Figure 2-4 (*a*) Period-2 motion for forced motion of a buckled beam in the phase plane (bending strain versus strain rate). (*b*) Chaotic trajectory for forced motion of a buckled beam.

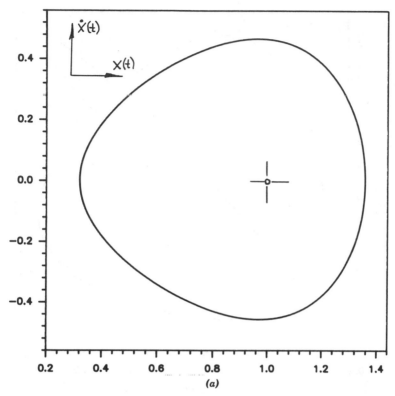

Figure 2-5 (*a*) Phase-plane trajectory of Duffing oscillator (1-2.4), $\alpha = -1$, $\beta = 1$.

properties similar to those in the classic phase plane $[x(t), \dot{x}(t)]$. In Figure 2-5 we show a pseudo-phase-plane orbit for a harmonic oscillator for different time delays. If the motion is chaotic, the trajectories do not close (Figure 2-6). The choice of T is not crucial, except to avoid a natural period of the system. When the state variables are greater than three (position, velocity, time or forcing phase), the higher-dimensional pseudo-phase-space trajectories can be constructed using multiple delays. For example, a three-dimensional space can be constructed using a vector with components $(x(t), x(t + T), x(t + 2T))$. More will be said about this technique in Chapter 4.

(e) Fourier Spectrum

One of the clues to detecting chaotic vibrations is the appearance of a broad spectrum of frequencies in the output when the input is a single-frequency harmonic motion or is dc (Figure 2-7). This characteristic of

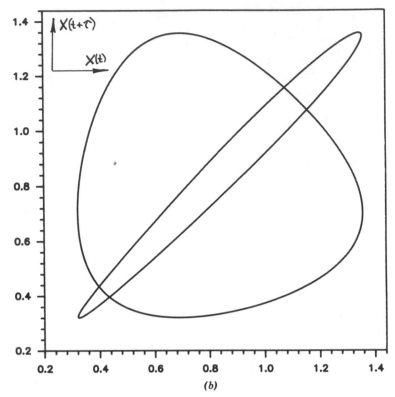

(b)

Figure 2-5 (*b*) Pseudo-phase-plane trajectory for the periodic oscillator in (*a*) for two delay times.

chaos becomes more important if the system is of low dimension (e.g., one to three degrees of freedom). Often, if there is an initial dominant frequency component ω_0, a precursor to chaos is the appearance of subharmonics ω_0/n in the frequency spectrum (see below). In addition to ω_0/n, harmonics of this frequency will also be present in the form $m\omega_0/n$ ($m, n = 1, 2, 3, \ldots$). An illustration of this test is shown in Figure 2-7*a–c*. The top figure shows a single spike in both the driving force and the response of a buckled beam. The middle spectrum shows a subharmonic motion and in the bottom figure a broad spectrum is shown, indicating possible chaotic motions.

One must be cautioned against concluding that multiharmonic outputs imply chaotic vibrations since the system in question might have many hidden degrees of freedom of which the observer is unaware. In large-degree-of-freedom systems, the use of the Fourier spectrum may not be of much help in detecting chaotic vibrations unless one can observe changes in

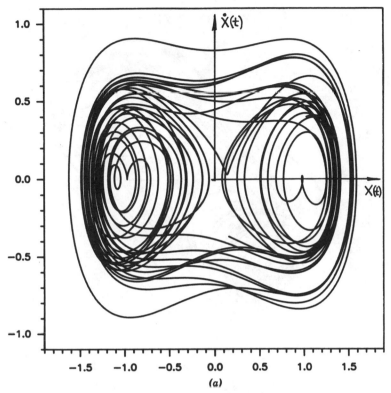

Figure 2-6 (*a*) Phase-plane trajectory for chaotic motion of a particle in a two-well potential (buckled beam) under periodic forcing (1-2.4), $\alpha = -1$, $\beta = 1$.

the spectrum as one varies some parameter such as driving amplitude or frequency.

(f) Poincaré Maps

In the mathematical study of dynamical systems a map refers to a time-sampled sequence of data $\{x(t_1), x(t_2), \ldots, x(t_n), \ldots x(t_N)\}$ with the notation $x_n \equiv x(t_n)$. A simple deterministic map is one in which the value of x_{n+1} can be determined from the values of x_n. This is often written in the form

$$x_{n+1} = f(x_n) \qquad (2\text{-}2)$$

This can be recognized as a *difference equation*. The idea of a map can be

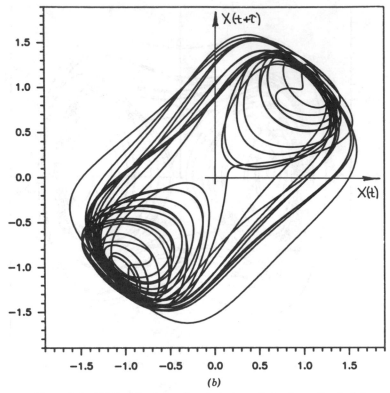

Figure 2-6 (*b*) Pseudo-phase-plane trajectory of chaotic motion in (*a*).

generalized to more than one variable. Thus, $\mathbf{x}_n$ could represent a vector with M components $\mathbf{x}_n = (Y1n, Y2n, \ldots, YMn)$ and Eq. (2-2) could represent a system of M equations.

For example, suppose we consider the motion of a particle as displayed in the phase plane $(x(t), \dot{x}(t))$. We learned above that when the motion is chaotic the trajectory tends to fill up a portion of phase space. However, if instead of looking at the motion continuously, we look only at the dynamics at discrete times, then the motion will appear as a sequence of dots in the phase plane (Figure 2-8). If $x_n \equiv x(t_n)$ and $y_n \equiv \dot{x}(t_n)$, this sequence of points in the phase plane represents a *two-dimensional* map

$$x_{n+1} = f(x_n, y_n)$$

$$y_{n+1} = g(x_n, y_n)$$

(2-3)

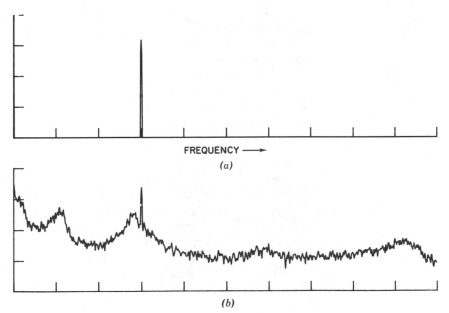

FREQUENCY ⟶

(a)

(b)

Figure 2-7 (a) Frequency spectrum of buckled elastic beam for low-amplitude excitation—linear periodic response. (b) Frequency spectrum of buckled elastic beam for larger excitation—broad-band response of beam due to chaotic vibration.

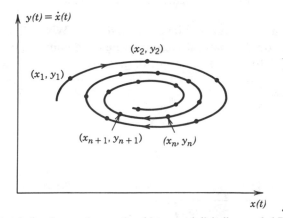

Figure 2-8 Sketch showing continuous time history and digitally sampled Poincaré points.

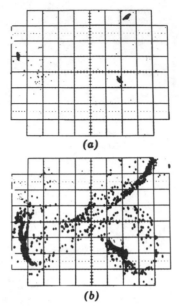

Figure 2-9 (*a*) Phase-plane Poincaré map showing a period-3 subharmonic motion of a periodically forced buckled beam. (*b*) Chaotic motion near a period-3 subharmonic.

When the sampling times t_n are chosen according to certain rules, to be discussed below, this map is called a *Poincaré map*.

Poincaré Maps for Forced Vibration Systems. When there is a driving motion of period T, a natural sampling rule for a Poincaré map is to choose $t_n = nT + \tau_0$. This allows one to distinguish between periodic motions and

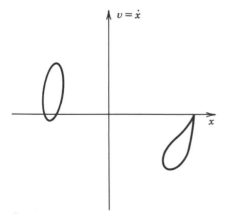

Figure 2-10 Phase-plane Poincaré map showing a quasiperiodic motion of a periodically forced two-degree-of-freedom beam in a two-well magnetic potential.

nonperiodic motions. For example, if the sampled harmonic motion shown in Figure 2-4*a*, is synchronized with its period, its "map" in the phase plane will be two points. If the output, however, were a subharmonic of period 3, the Poincaré map would consist of a set of three points as shown in Figure 2-9a. [In the jargon of mathematical dynamics, one says that the functions $f(\ )$ and $g(\ )$ in Eq. (2.3) have three *fixed points*.]

Another nonchaotic Poincaré map is shown in Figure 2-10, where the motion consists of two *incommensurate* frequencies

$$x(t) = C_1 \sin(\omega_1 t + d_1) + C_2 \sin(\omega_2 t + d_2) \tag{2-4}$$

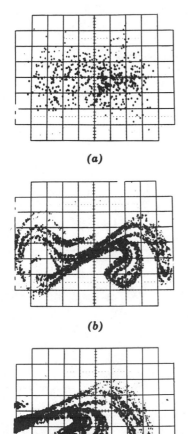

(a)

(b)

(c)

Figure 2-11 (*a*) Poincaré map of chaotic motion of a buckled beam with low damping. (*b*, *c*) Poincaré map of chaotic motion of a buckled beam for higher damping showing fractal-like structure of a strange attractor [from Moon (1980a) with permission of ASME, copyright 1980].

where ω_1/ω_2 is an irrational number. If one samples at a period corresponding to either frequency, the map in the phase plane will become a continuous closed figure or orbit. This motion is sometimes called *almost periodic* or *quasiperiodic* motion or "motion on a torus" and is not considered to be chaotic.

Finally, if the Poincaré map does not consists of either a finite set of points (Figure 2-9a) or a closed orbit (Figure 2-10), the motion may be chaotic (Figure 2-11). Here we must distinguish between damped and undamped systems. In undamped or lightly damped systems, the Poincaré maps of chaotic motions often appear as a cloud of unorganized points in the phase plane (Figure 2-11a). Such motions are sometimes called *stochas-*

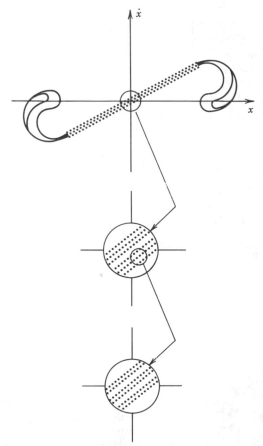

Figure 2-12 Poincaré map of chaotic vibration of a forced nonlinear oscillator showing self-similar structure at finer and finer scales.

TABLE 2-2   Classification of Poincaré Maps

Finite number of points: periodic or subharmonic oscillation
Closed curve: quasiperiodic, two incommensurate frequencies present
Open curve: suggest modeling as a one-dimensional map; try plotting $x(t)$ versus $x(t + T)$
Fractal collection of points: strange attractor in three phase space dimensions
Fuzzy collection of points: (i) dynamical system with too much random or noisy input; (ii) strange attractor but system has very small dissipation—use Lyapunov exponent test; (iii) strange attractor in phase space with more than three dimensions—try multiple Poincaré map; (iv) quasiperiodic motion with three or more dominant incommensurate frequencies.

tic (e.g., see Lichtenberg and Lieberman, 1983). In damped systems, the Poincaré map will sometimes appear as an infinite set of highly organized points arranged in what appear to be parallel lines, as shown in Figure 2-11b,c. In numerical simulations, one can enlarge a portion of the Poincaré map (see Figure 2-12) and observe further structure. If this structured set of points continues to exist after several enlargements, one says that the motion behaves as a *strange attractor*. This embedding of structure within structure is often referred to as a *Cantor set* (see Chapter 6).

The appearance of Cantor set-like patterns in the Poincaré map of a vibration history is a strong indicator of chaotic motions. The classes of patterns of Poincaré maps are listed in Table 2-2.

Poincaré Maps in Autonomous Systems. Steady-state vibrations can also be generated without periodic or random inputs if the motion originates from a dynamic instability such as wind-induced flutter in an elastic structure (Figure 2-13) or a temperature-gradient-induced convection motion in a fluid or gas (e.g., Benard convection, Figure 1-23). In electrical systems or feedback control devices, self-excited oscillations can arise from negative resistance elements or negative feedback. One is then led to ask how to choose the sampling times in a Poincaré map. Here the discussion gets a little abstract.

Consider the lowest-order chaotic system governed by three first-order differential equations (e.g., the Lorenz equations of Chapter 1). In an electromechanical system, the variables $x(t)$, $y(t)$, and $z(t)$ could represent displacement, velocity, and control force as in a feedback-controlled system. We then imagine the motion as a trajectory in a three-dimensional phase space (Figure 2-14). A Poincaré map can be defined by constructing a two-dimensional oriented surface in this space and looking at the points (x_n, y_n, z_n) where the trajectory pierces this surface. For example, we can

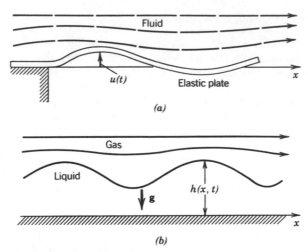

(a)

(b)

Figure 2-13 Examples of self-excited vibrations: (*a*) fluid flow over an elastic plate and (*b*) gas flow over a liquid interface.

choose a plane $n_1 x + n_2 y + n_3 z = c$ with normal vector $\mathbf{n} \equiv (n_1, n_2, n_3)$. As a special case, choose points where $x = 0$. Then the Poincaré map consists of points that pierce this plane with the same sense; that is, if $\mathbf{s}(t)$ represents a unit vector along the trajectory, $\mathbf{s}(t_n) \cdot \mathbf{n}$ must always have the same sign.

This definition of the Poincaré map actually includes the case when the system is periodically forced. Consider, for example, a forced nonlinear

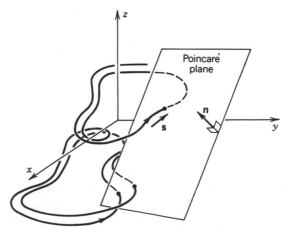

Figure 2-14 Sketch of time evolution trajectories of a third-order system of equations and a typical Poincaré plane.

oscillator with equations of motion

$$\dot{x} = y \tag{2-5}$$

$$\dot{y} = F(x, y) + f_0\cos(\omega t + \phi_0) \tag{2-6}$$

This system can be made to look like an autonomous one by defining

$$z = \omega t + \phi_0 + 2n\pi \tag{2-7}$$

and

$$\dot{x} = y \tag{2-8}$$

$$\dot{y} = F(x, y) + f_0\cos z \tag{2-9}$$

$$\dot{z} = \omega \tag{2-10}$$

Thus, a natural sampling time is chosen when $z = 0$. This system can be thought of as a cylindrical phase space where the values of z are restricted: $0 \le z \le 2\pi$. A picture of the Poincaré map is then given as in Figure 2-15.

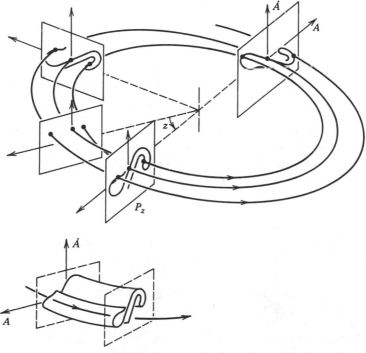

Figure 2-15 Sketch of a strange attractor for a forced nonlinear oscillator: product space of the Poincaré plane and the phase of the forcing excitation.

Reduction of Dynamics to One-Dimensional Maps.   In Chapter 1 we saw
that simple one-dimensional maps or difference equations of the form
$x_{n+1} = f(x_n)$ can exhibit period-doubling bifurcations and chaos when the
function $f(x)$ has at least one maximum (or minimum), as in Figure 1-19.
Period-doubling phenomena have been observed in so many different
complex physical systems (fluids, lasers, $p-n$ electronic junctions) that in
many cases the dynamics may sometimes be modeled as a one-dimensional
map. This is especially possible in systems with significant dissipation. To
check this possibility, one samples some dynamic variable using a Poincaré
section as discussed above; that is, $x_n = x(t = t_n)$. Then one plots each x_n
against its successor value x_{n+1}. Two criteria must be met to declare the
system chaotic. First, the points x_{n+1} versus x_n must appear to be clustered
in some apparent functional relationship and second, this function $f(x)$
must be multivalued, that is, it must have a maximum or a minimum. If this
be the case, one then attempts to fit a polynomial function to the data and
uses this mapping to do numerical experiments or analysis along the lines
of the quadratic map (Chapters 1 and 5). Examples of this technique may
be found in Shaw (1984) in the problem of a dripping faucet and in Rollins
and Hunt (1982) in an experiment with a varactor diode in an electrical
circuit (see also Chapter 3 for a discussion of these problems). This
technique is discussed further in Chapter 4.

(g)   Routes to Chaos

Periodic to Chaotic Motions Through Parameter Changes.   In conducting
any of these tests for chaotic vibrations, one should try to vary one or more
of the control parameters in the system. For example, in the case of the
buckled structure (Figure 2-2), one can vary either the forcing amplitude or
forcing frequency, or in the case of a nonlinear circuit, one can vary the
resistance. The reason for this procedure is to see if the system has steady
or periodic behavior for some range of the parameter space. In this way,
one can have confidence that the system is in fact deterministic and that
there are no hidden inputs or sources of truly random noise.
    In changing a parameter, one looks for a pattern of periodic responses.
One characteristic precursor to chaotic motion is the appearance of subharmonic periodic vibrations. There may in fact be many patterns of
prechaos behavior. Several models of prechaotic behavior have been observed in both numerical and physical experiments. (See e.g. Gollub and
Benson, 1980, or Kadanoff, 1983.)

Period-Doubling Route to Chaos.   In the period-doubling phenomenon,
one starts with a system with a fundamental periodic motion. Then as some

experimental parameter is varied, say λ, the motion undergoes a bifurcation or change to a periodic motion with twice the period of the original oscillation. As λ is changed further, the system bifurcates to periodic motions with twice the period of the previous oscillation. One outstanding feature of this scenario is that the critical values of λ at which successive period doublings occur obey the following scaling rule (see also Chapter 1):

$$\frac{\lambda_n - \lambda_{n-1}}{\lambda_{n+1} - \lambda_n} \rightarrow \delta = 4.6692016 \qquad (2\text{-}11)$$

as $n \rightarrow \infty$. (This is called the Feigenbaum number after the physicist who discovered this scaling behavior.) In practice, this limit approaches δ by the third or fourth bifurcation.

This process will accumulate at a critical value of the parameter, after which the motion becomes chaotic.

This phenomenon has been observed in a number of physical systems as well as numerical simulations. The most elementary mathematical equation that illustrates this behavior is a first-order difference equation (see Chapter 1):

$$x_{n+1} = 4\lambda x_n (1 - x_n) \qquad (2\text{-}12)$$

As the system parameter is changed beyond the critical value, chaotic motions exist in a band of parameter values. However, these bands may be of finite width; that is, as the parameter is varied, periodic windows may develop. Periodic motions in this regime may again undergo period-doubling bifurcations leading once more to chaotic motions (see §5.3).

The period-doubling model for the route to chaos is an elegant, aesthetic model and has been described in many popular articles. However, while many physical systems exhibit properties similar to those of (2-12), many other systems do not. Nevertheless, when chaotic vibrations are suspected in a system, it is worthwhile checking to see if period doubling is present.

Bifurcation Diagrams. A widely used technique for examining the pre-chaotic or postchaotic changes in a dynamical system under parameter variations is the bifurcation diagram (an example is shown in Figure 2-16). Here some measure of the motion (e.g., maximum amplitude) is plotted as a function of a system parameter such as forcing amplitude or damping constant. If the data are sampled using a Poincaré map, it is very easy to observe period doubling and subharmonic bifurcations as shown in the experimental data for a nonlinear circuit from a paper by Bryant and Jeffries (1984a, b) at the University of California, Berkeley. However, when

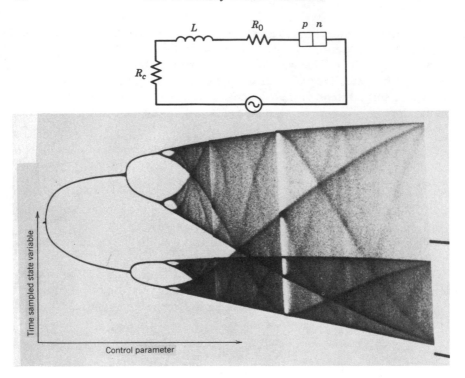

Figure 2-16 Experimental bifurcation diagram for a periodically forced nonlinear circuit with a $p - n$ junction; periodically sampled current versus drive amplitude voltage [from Van Buskirk and Jeffries (1985) with permission of The American Physical Society, copyright 1985].

the bifurcation diagram loses continuity, it may mean either quasiperiodic motion or chaotic motion and further tests are required to classify the dynamics.

Quasi-Periodic Route to Chaos.   While period doubling is the most celebrated scenario for chaotic vibration, there are several other schemes that have been studied and observed. In one proposed by Newhouse et al. (1978), they imagine a system which undergoes successive dynamic instabilities before chaos. For example, suppose a system is initially in a steady state and becomes dynamically unstable after changing some parameter (e.g., flutter). As the motion grows, nonlinearities come into effect and the motion becomes a limit cycle. Such transitions are called *Hopf bifurcations* in mathematics (e.g., see Abraham and Shaw, 1983). If after further parameter changes the system undergoes two more Hopf bifurcations so

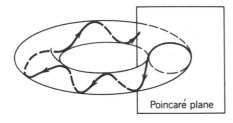

Poincaré plane

Figure 2-17 Sketch illustrating the coupled motion of two oscillators and the Poincaré plane used to detect a quasiperiodic route to chaos.

that three simultaneous coupled limit cycles are present, then chaotic motions become possible.

Thus, the precursor to such chaotic motion is the presence of two simultaneous periodic oscillations. When the frequencies of these oscillations, ω_1 and ω_2, are not commensurate, the observed motion itself is not periodic but is said to be *quasiperiodic* [see Eq. (2-4)]. As discussed above, the Poincaré map of a quasiperiodic motion is a closed curve in the phase plane (Figure 2-10). Such motions are imagined to take place on the surface of a torus where the Poincaré map represents a plane which cuts the torus (see Figure 2-17). If ω_1 and ω_2 are incommensurate, the trajectories fill the surface of the torus. If ω_1/ω_2 is a rational number, the trajectory on the torus will eventually close although it might perform many orbits in both angular directions of the torus before closing. In the latter case, the Poincaré map will become a set of points generally arranged in a circle. Chaotic motions are often characterized in such systems by the breakup of the quasiperiodic torus structure as the system parameter is varied (Figure 2-18).

Evidence for the three-frequency transition to chaos has been observed in flow between rotating cylinders (Taylor–Couette flow) where vortices form with changes in the rotation speed. Three Fourier spectra from one such experiment are shown in Figure 2-19. In the top figure one periodic motion appears to be present and in the second two major motions are evident. In the third we have the sign of an increase in broad-band noise which is characteristic of chaotic behavior.

Intermittency. In a third route to chaos, one observes long periods of periodic motion with bursts of chaos. This scenario is called *intermittency*. As one varies a parameter, the chaotic bursts become more frequent and longer (e.g., see Manneville and Pomeau, 1980). Evidence for this model for prechaos has been claimed in experiments on convection in a cell (closed box) with a temperature gradient (called Rayleigh–Bénard convection) (see Figure 2-20). Some models for intermittency predict that the average time

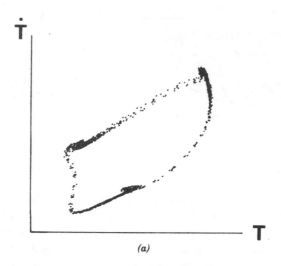

(a)

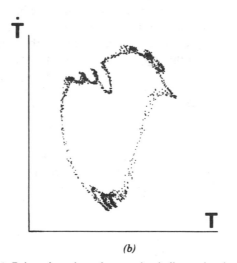

(b)

Figure 2-18 (*a*) Poincaré section of a quasiperiodic motion in Rayleigh–Benard thermal convection with a frequency ratio close to $\omega_1/\omega_2 = 2.99$. (*b*) Breakup of the torus surface prior to the onset of chaos [from Berge (1982)].

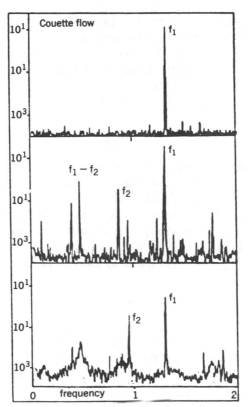

Figure 2-19 Evidence for the three-frequency transition to chaos in the flow between rotating cylinders (Taylor–Couette flow); the rotation difference increases from top to bottom [from Swinney and Gullub (1978)].

Figure 2-20 Sketch of the time history for intermittent-type chaos.

of the regular or laminar phase of the motion $\langle \tau \rangle$ will scale in a precise way as some system parameter is varied; for example,

$$\langle \tau \rangle \approx \frac{1}{(\lambda - \lambda_c)^{1/2}} \qquad (2\text{-}13)$$

where λ_c is the value at which the periodic motion becomes chaotic.

It should be noted that for some physical systems one may observe all three patterns of prechaotic oscillations and many more depending on the parameters of the problem. The benefit in identifying a particular prechaos pattern of motion with one of these now "classic" models is that a body of mathematical work on each exists which may offer better understanding of the chaotic physical phenomenon under study.

Crisis

One sign of the revolutionary ideas coming into the field of dynamics and vibrations is the welter of new names for new phenomena. One new word coined by Grebogi et al. (1983a) is the term *crisis*, which is used to denote a sudden change in the chaotic state when some system parameter is changed. For example, a system initially in a chaotic state may suddenly become periodic. Or chaotic motion which was originally confined to a limited range of $x(t)$ may suddenly expand to a broad range $x(t)$.

Transient Chaos

Sometimes chaotic vibrations appear for some parameter changes but eventually settle into a periodic or quasiperiodic motion after a short time. According to Grebogi et al. (1983b), such transient chaos is a consequence of a crisis or the sudden disappearance of sustained chaotic dynamics. Thus, experiments and numerical simulation should be allowed to run for a time after one thinks the system is in chaos even if the Poincaré map seems to be mapping out a fractal structure characteristic of strange attractors. How long should one wait before calling the state chaotic? Right now this is a judgment call. In our laboratory we require 4000 Poincaré points in a fractal-looking attractor before pronouncing the state chaotic. See Chapter 5 for further discussion of transient chaos.

Conservative Chaos

While much of the new excitement about nonlinear dynamics has focused on chaotic dynamics in dissipative systems, chaotic behavior in nondissipative or so-called conservative systems had been known for some time. In fact, the search for solutions to the equations of celestial mechanics in the late 19th century led mathematicians like Poincaré to speculate that many dynamic problems were sensitive to initial conditions and hence were unpredictable in the details of the motions of orbiting bodies.

The study of chaotic dynamics in energy-conserving systems, while not the principal focus of this book, has received much attention in the literature and sometimes is found under the heading of "Hamiltonian Dynamics" which refers to the methods of Hamilton (and also Jacobi) that are used to solve nonlinear problems in multi-degree-of-freedom nondissipative systems (e.g., see the excellent monograph by Lichtenberg and Lieberman, 1983).

Our purpose here is to give a purely descriptive picture of chaos in such systems to contrast the properties of nonpredictable dynamics in both nonconservative and conservative problems.

Examples of conservative systems in the physical world include orbital problems in celestial mechanics and the behavior of particles in electromagnetic fields. Hence, much of the work in this field has been done by those interested in plasma physics, astronomy, and astrophysics.

However, while most earth-bound dynamics problems have some energy loss, some, like structural systems or microwave cavities, have very little damping and over a finite period of time can behave like a conservative or Hamiltonian system. An example might be the vibration of an orbiting space structure. Also, conservative system dynamics provides a limiting case for small damping dynamic analysis. Thus, while we do not attempt to present a rigorous or lengthy summary of Hamiltonian dynamics, it is useful to discuss the general features of these problems.

Typically, energy-conserving systems can exhibit the same types of bounded vibratory motion as lossy systems including periodic, subharmonic, quasiperiodic, and chaotic motions. One of the main differences, however, between vibrations in lossy and lossless problems is that chaotic orbits in lossy systems exhibit a fractal structure in the phase space whereas chaotic orbits in lossless systems do not.

Chaotic orbits in conservative systems tend to visit all parts of a subspace of the phase space uniformly; that is, they exhibit a uniform probability density over restricted regions in the phase space. Thus, lossless systems exhibit different Poincaré maps from those of lossy problems. However, the use of Lyapunov exponents as a measure of nearby orbit divergence is still useful. An example of a system with no dissipation is the ball bouncing on an elastic table where the table is moving and the impact is assumed to be lossless or elastic. Details of this problem are discussed in Chapter 5.

Lyapunov Exponents and Fractal Dimensions

The tests for chaotic vibrations described in this chapter are mainly qualitative and involve some judgment and experience on the part of the investigator. Quantitative tests for chaos are available and have been used with some success. Two of the most widely used criteria are the Lyapunov exponent (see Chapter 5) and the fractal dimension (see Chapter 6). In summary, these two indicators are currently interpreted as follows:

1. Positive Lyapunov exponents imply chaotic dynamics.
2. Fractal dimension of the orbit in phase space implies the existence of a strange attractor.

The Lyapunov exponent test can be used for both dissipative or nondissipative (conservative) systems, while the fractal dimension test only makes sense for dissipative systems.

The Lyapunov exponent test measures the sensitivity of the system to changes in initial conditions. Conceptually, one imagines a small ball of initial conditions in phase space and looks at its deformation into an ellipsoid under the dynamics of the system. If d is the maximum length of the ellipsoid and d_0 the initial size of the initial condition sphere, the Lyapunov exponent λ is interpreted by the equation

$$d = d_0 2^{\lambda(t - t_0)}$$

One measurement, however, is not sufficient and the calculation must be averaged over different regions of phase space. This average can be represented by

$$\lambda = \lim_{N \to \infty} \frac{1}{N} \sum_{1}^{N} \frac{1}{(t_i - t_{0i})} \ln_2 \frac{d_i}{d_{0i}}$$

A more detailed discussion is given in Chapter 5 along with references.

The fractal dimension is related to the discussion of the horseshoe map in Chapter 1. There we saw that in a chaotic dynamic system regions of phase space are stretched, contracted, folded, and remapped onto the original space. This remapping for dissipative systems leaves gaps in the phase space. This means that orbits tend to fill up less than an integer subspace in phase space. The fractal dimension is a measure of the extent to which orbits fill a certain subspace and a noninteger dimension is a hallmark of a *strange attractor*. There are many definitions of fractal dimension but the most basic one is derived from the notion of counting the number of spheres N of size ϵ needed to cover the orbit in phase space. Basically, $N(\epsilon)$ depends on the subspace of the orbit. If it is a periodic or limit cycle orbit then $N(\epsilon) \approx \epsilon^{-1}$. When the motion lies on a strange attractor, $N(\epsilon) \approx \epsilon^{-d}$ or

$$d = \lim_{\substack{N \to \infty \\ \epsilon \to 0}} \frac{\ln N}{\ln(1/\epsilon)}$$

Further discussion is given in Chapter 6.

While both quantitative tests can be automated using computer control, experience and judgment are still required to provide a conclusive assessment as to whether the motion is *chaotic* or *strange attractor*. Finally, almost all physical examples of *strange attractors* have been found to be chaotic, that is, noninteger d implies $\lambda > 0$. However, a few mathematical models have been studied where one does not imply the other (e.g., see Grebogi et al., 1984).

3

A Survey of Systems with Chaotic Vibrations

*The world is what it is and I am what I am.... This out
there and this in me, all this, everything, the resultant of
inexplicable forces. A chaos whose order is beyond
comprehension. Beyond human comprehension.*

Henry Miller, *Black Spring*

3.1 NEW PARADIGMS IN DYNAMICS

In his book *The Structure of Scientific Revolutions*, Thomas Kuhn (1962) argues that major changes in science occur not so much when new theories are advanced but when the simple models with which scientists conceptualize a theory are changed. A conceptual model or problem that embodies the major features of a whole class of problems is called a *paradigm*. In vibrations, the spring–mass model represents such a paradigm. In nonlinear dynamics the motion of the pendulum and the three-body problem in celestial mechanics represent classical paradigms.

The theory that new models or paradigms are precursors for major changes in scientific or mathematical thinking has no better example than the current revolution in nonlinear dynamics. Here the two principal paradigms are the Lorenz attractor [Eq. (1-3.9)] and the logistic equation [Eq. (1-3.6)]. Many of the features of chaotic dynamics are embodied in these two examples, such as divergent trajectories, subharmonic bifurcations, period doubling, Poincaré maps, and fractal dimensions. Just as the novitiate in linear vibrations had to master the subtleties of the spring–mass paradigm to understand vibrations of complex systems, so the budding nonlinear dynamicist of today must understand the phenomena embedded in the Lorenz and logistic models. Other lesser paradigms are also important in dynamical systems, including the forced motions of the Van der Pol equation (1-2.5), the Duffing oscillator models (1-2.4) of Ueda and Holmes (see below, this chapter), and the two-dimensional map of Henon (1-3.8).

Readers interested in a detailed discussion of the Lorenz model for thermal convection in a fluid should read Sparrow's monograph (1982) on the subject. Guckenheimer and Holmes (1983) have written an advanced mathematical text based on four paradigms of modern dynamics—Van der Pol's equation, the Duffing model for a buckled beam, the Lorenz equation, and the Henon attractor. Another classic model for chaotic dynamics is a mass subject to impact forces such as a ball bouncing on a vibrating table or bouncing between two walls. This model is also applicable to acceleration of electrons in electromagnetic fields and is sometimes called the Fermi acceleration model. This model results in a two-dimensional map similar to that of Henon. A good discussion of the Fermi model as well as the Lorenz equation may be found in Lichtenberg and Lieberman (1983).

Most of the above-mentioned texts, however, focus almost entirely on the mathematical analysis of these models for chaos. In this chapter, we survey a variety of mathematical and physical models which exhibit chaotic vibrations. An attempt is made to describe the physical nature of the chaos in these examples and to point out the similarities and differences between the physical problems and their more mathematical chaos paradigms mentioned above. These examples are drawn from solid and fluid mechanics, electric circuits, control theory, and chemical engineering. Some of the experimental evidence obtained to date for the existence of chaotic vibrations is highlighted.

3.2 MATHEMATICAL MODELS OF CHAOTIC PHYSICAL SYSTEMS

We distinguish here between mathematical *models* derived from physics that exhibit chaotic dynamics and physical *experiments* in which chaotic motions have actually been observed. Readers with access to a small computer can observe chaotic solutions of many of these models by using a Runge–Kutta numerical integration scheme. Sample problems with suggested parameters for a few of these models are given in Appendix B.

Thermal Convection in Fluids

Perhaps the most famous model to date is the Lorenz equations which attempt to model atmospheric dynamics. In this model, one imagines a fluid

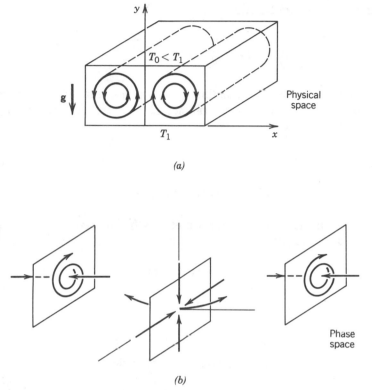

Figure 3-1 (*a*) Sketch of thermofluid convection rolls. (*b*) Three unstable singular points in phase space for the Lorenz equations (3-2.3).

layer, under gravity, which is heated from below so that a temperature difference is maintained across the layer (Figure 3-1). When this temperature difference becomes large enough, circulatory, vortex-like motion of the fluid results in which the warm air rises and the cool air falls. The tops of parallel rows of convection rolls can sometimes be seen when flying above a cloud layer. The two-dimensional convective flow is assumed to be governed by the classic Navier–Stokes equations (1-1.3). These equations are expanded in the two spatial directions in Fourier modes with fixed boundary conditions on the top and bottom of the fluid layer. For a small temperature difference ΔT, no fluid motion takes place, but at a critical ΔT, convective or circulation flow occurs. This motion is referred to as *Rayleigh–Benard convection*.

Truncation of the Fourier expansion in three modes was studied by Lorenz (1963). An earlier study by Saltzman (1962) used a five-mode truncation. In this simplification, the velocity in the fluid (v_x, v_y) is written in terms of a stream function ψ:

$$v_x = \frac{\partial \psi}{\partial y} \qquad v_y = \frac{-\partial \psi}{\partial x}$$

In the Lorenz model, the nondimensional stream function and perturbed temperature are written in the form (see Lichtenberg and Lieberman, 1983, pp. 443–446 for a derivation)

$$\psi = \sqrt{2}\, x(t)\sin \pi a x \sin \pi y \qquad (3\text{-}2.1)$$

$$\theta = \sqrt{2}\, y(t)\cos \pi a x \sin \pi y - z(t)\sin 2\pi y \qquad (3\text{-}2.2)$$

where the fluid layer is taken as a unit length. The resulting equations for (x, y, z) are then given by

$$\dot{x} = \sigma(y - x)$$

$$\dot{y} = \rho x - y - xz \qquad (3\text{-}2.3)$$

$$\dot{z} = -\beta z + xy$$

The parameter σ is a nondimensional ratio of viscosity to thermal conductivity (Prandtl number), ρ is a nondimensional temperature gradient (related to the Rayleigh number), and $\beta = 4(1 + a^2)^{-1}$ is a geometric factor, with $a^2 = \frac{1}{2}$.

For the parameter values $\sigma = 10$, $\rho = 28$, and $\beta = \frac{8}{3}$ (studied by Lorenz), there are three equilibrium points, all of them unstable (Figure 3-1b). The origin is a saddle point, while the other two are unstable foci or spiral equilibrium points (see Figure 1-24). However, globally, one can show that the motion is bounded. Thus, the trajectories have no home but remain confined to an ellipsoidal region of phase space. A numerical example of one of these wandering trajectories is shown in Figure 1-25.

Thermal Convection Model of Moore and Spiegel

It is often the case that discoveries of major significance are not singular and that different people in different places observe new phenomena at about the same time. Such appears to be the case regarding the discovery of low-order models for thermal convection dynamics. Above we discussed the now famous Lorenz (1963) equations, (3-2.3), which later received tremendous attention from mathematicians. Yet around the same time,

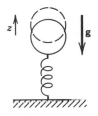

Figure 3-2 (*a*) Spring–mass model for thermal convection of Moore and Spiegel (1966).

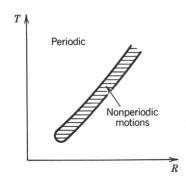

Figure 3-2 (*b*) Region of nonperiodic motions in the nondimensional parameter space for the thermal convection model of Moore and Spiegel (1966), Eq. (3-2.4).

Moore and Spiegel (1966) of the Goddard Institute and New York University, respectively, proposed a model for unstable oscillations in fluids which rotate, have magnetic fields or are compressible, and have thermal dissipation. The equations derived in their paper, like Lorenz's, are equivalent to three first-order differential equations. If z represents the vertical displacement of a compressible fluid mass in a horizontally stratified fluid (Figure 3-2*a*), restoring forces in the fluid are represented by a spring force and a buoyancy force resulting from gravity. Also, the fluid element can exchange heat with the surrounding fluid. Thus, the dynamics is modeled by a second-order equation (Newton's law) coupled to a first-order equation for heat transfer, leading to a third-order equation. In nondimensional form this equation becomes

$$\dddot{z} + \ddot{z} + (T - R + Rz^2)\dot{z} + Tz = 0 \qquad (3\text{-}2.4)$$

where a nonlinear temperature profile of the form

$$\theta = \theta_0\left[1 - \left(\frac{z}{L}\right)^2\right]$$

is assumed. In Eq. (3-2.4), T and R are nondimensional groups:

$$T = \left(\frac{\text{thermal relaxation time}}{\text{free oscillation time}} \right)^2$$

$$R = \left(\frac{\text{thermal relaxation time}}{\text{free fall time}} \right)^2$$

In their numerical studies, Moore and Spiegel discovered an entire region of *aperiodic* motion as shown in Figure 3-2b. In a follow-up paper, Baker et al. (1971) analyzed the stability of periodic solutions in the aperiodic regime.

They showed that Eq. (3-2.4) can be put into the form

$$\ddot{s} = -(1 - \delta)s + \theta$$

$$\dot{\theta} = -R^{-1/2}\theta + (1 - \delta s^2)\dot{s}$$

(3-2.5)

The limit of $R \to \infty$ is the zero dissipative case. In this limit (R large), Baker et al. show that in the range of periodicity, the periodic solutions of (3-2.6) become unstable locally. This property of global stability and local instability seems to be characteristic of chaotic differential equations. In a more recent paper, Marzec and Spiegel (1980) studied a more general class of third-order equations of the form

$$\dot{x} = y$$

$$\dot{y} = -\frac{dV(x, \lambda)}{dx} - \epsilon\mu y$$

(3-2.6)

$$\dot{\lambda} = -\epsilon[\lambda + g(x)]$$

where $V(x, \lambda)$ is thought of as a potential function. They show that both the Moore–Spiegel oscillator (3-2.4) and the Lorenz system (3-2.3) (with a change of variables) can be put into the above form (3-2.6). Strange attractor solutions to specific examples of (3-2.6) were found numerically.

The above set of equations also models a second-order oscillator with feedback control λ.

It will be an interesting study for historians of science to answer why the Lorenz system received so much study and the Moore–Spiegel model was virtually ignored by mathematicians. Both purported to model convection. Lorenz published his article in the *Journal of Atmospheric Sciences* while Moore and Spiegel published theirs in the *Astrophysics Journal*.

Supersonic Panel Flutter

An analytic–analog computer study that uncovered chaotic vibrations and predates the Lorenz paper by one year is that of Kobayashi (1962). He analyzed the vibrations of a buckled plate with supersonic flow on one side of the plate. This problem, known as "panel flutter," was important not only for supersonic aircraft, but for the then emerging rocket technology. Kobayashi expanded the deflection of the simply supported plate in a Fourier series and studied the coupled motion of the first two modes. Denoting the nondimensional modal amplitudes of these two modes by x and y, the equations studied using an analog computer were of the form

$$\ddot{x} + \delta\dot{x} + \left[1 - q + x^2 + 4y^2\right]x - Qy = 0$$
$$\ddot{y} + \delta\dot{y} + 4\left[4 - q + x^2 + 4y^2\right]y + Qx = 0 \tag{3-2.7}$$

where q is a measure of the in-plane compressive stress in the plate (which can exceed the buckling value) and Q is proportional to the dynamic fluid pressure of the supersonic flow upstream of the plate. In his abstract of this 1962 paper Kobayashi states, "Moreover the following remarkable results are obtained. (i) In some unstable region of a moderately buckled plate, only an *irregular vibration* is observed" [italics added]. He also refers to earlier experimental studies in 1957 at the NACA in the United States which was the pre-Sputnik ancestor of NASA (see also Fung, 1958).

On reading some of this early literature, it is clear that chaotic vibration has been observed in the past but that no models for its analysis were available at the time.

Impact Force Problems

Impact-type problems result in explicit difference equations or maps which often yield chaotic vibration under certain parameter conditions. A classic impact-type map is the motion of a particle between two walls. When one wall is stationary and the other oscillatory (Figure 3-3*a*), the problem is called the Fermi model for cosmic ray acceleration involving charged particles and moving magnetic fields. This model is discussed in great detail by Lichtenberg and Lieberman (1983) in their readable monograph on stochastic motion. Several sets of difference equations have been studied for this model. One set in which one wall imparts momentum changes without change of position is given by

$$v_{n+1} = |v_n + V_0\sin \omega t_n|$$
$$t_{n+1} = t_n + \frac{2\Delta}{v_{n+1}} \tag{3-2.8}$$

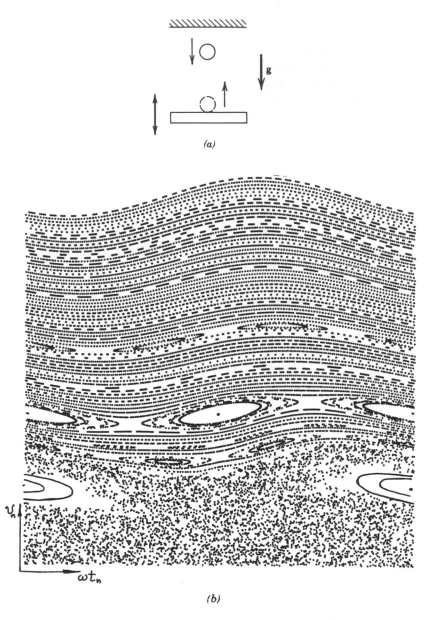

Figure 3-3 (*a*) Particle impact dynamics model with a periodically vibrating wall. (*b*) Poincaré map v_n *vs* $\omega t_n \, (\mathrm{mod}\, 2\pi)$ for the impact problem in (*a*) using Eqs. (3-2.8).

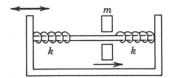

Figure 3-4 Experimental model of mass with a dead-band in the restoring force.

where v_n is the velocity after impact, t_n is the time of impact, V_0 is the maximum momentum per unit mass that the wall can impart, and Δ is the gap between the two walls.

Numerical studies of this and similar equations reveal that stochastic-type solutions exist in which thousands of iterations of the map (3-2.8) fill up regions of the *phase space* (v_n, t_n) as illustrated in Figure 3-3b. In some cases, the trajectory does not penetrate certain "islands" in the (v_n, t_n) plane. In these islands more regular orbits occur. This system can often be analyzed using classical Hamiltonian dynamics. This system is typical of chaos in low or zero dissipation problems. In moderate to high dissipation, the chaotic Poincaré map becomes localized in a structure with fractal properties as in Figure 2-11b, c. But in low dissipation problems, the Poincaré map fills up large areas of the phase plane with no apparent fractal structure.

The Fermi accelerator model is also similar to one in mechanical devices in which there exists play, as illustrated in Figure 3-4. A mass slides freely on a shaft with viscous damping until it hits stiff springs on either side (see Shaw and Holmes, 1983; Shaw, 1985a, b). Another mathematical model which is closer to the physics is the bouncing ball on a vibrating surface shown in Figure 3-5. This problem has been studied by Holmes (1982). Using an energy loss assumption for each impact, one can show that the following difference equations result:

$$\phi_{j+1} = \phi_j + v_j$$
$$v_{j+1} = \alpha v_j - \gamma \cos\left(\phi_j + v_j\right)$$

(3-2.9)

Here ϕ represents a nondimensional impact time, and v represents the velocity after impact. As shown in Figure 3-5a, a steady sinusoidal motion of the table can result in a nonperiodic motion of the ball. A fractual-looking chaotic orbit for this map is shown in Figure 3-5b.

Experiments on the chaotic boucing ball have been performed by Tufillaro and Albano (1986). Other studies of impact or *bilinear* oscillator problems have been done by Thompson and Ghaffari (1982), Thompson (1983) and Isomaki et al. (1985).

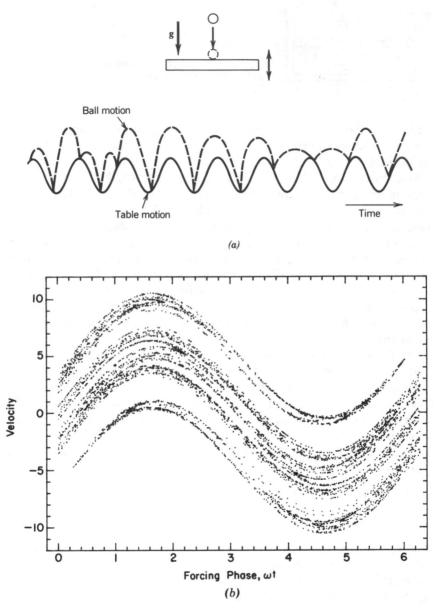

Figure 3-5 (a) Chaotic time history of a ball bouncing on a periodically vibrating table (Holmes, 1982). (b) Poincaré map of (a); Impact velocity versus impact time (modulus 2π).

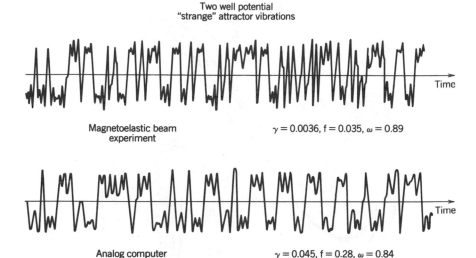

Figure 3-6 Chaotic vibrations of a periodically forced buckled beam: comparison of analog computer simulation and experimental measurements [from Moon and Holmes (1979)].

Double-Well Potential Problems

The forced vibrations of a buckled beam were modeled by a Duffing-type equation by Holmes (1979) who showed in analog computer studies that chaotic vibrations were possible. The nondimensional equation derived by Holmes is

$$\ddot{x} + \gamma\dot{x} - \tfrac{1}{2}x(1 - x^2) = f_0\cos \omega t \qquad (3\text{-}2.10)$$

where x represents the lateral motion of the beam (here a simple one-mode model is used to represent the beam). This equation can also model a particle in a two-well potential (Figure 1-3). This model has been used to study plasma oscillations (e.g., see Mahaffey, 1976). Chaotic solutions obtained from an analog computer are shown in Figure 3-6. An experimental realization of this model was discussed in Chapter 2. A Fourier spectrum based on solutions to this equation (Figure 2-7) shows a continuous spectrum of frequencies which is characteristic of chaotic motions. A Poincaré map of the strange attractor is shown in Figure 3-7. Fractal dimensions for chaotic solutions are discussed in later chapters. Numerical studies of the double-well problem have also been published by Dowell and Pezeshki (1986), Moon and Li (1985a, b), and Ueda et al. (1986).

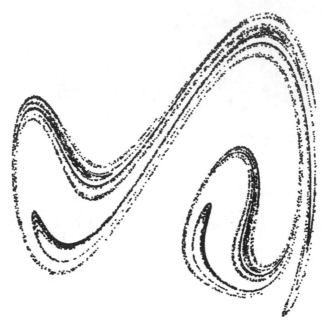

·**Figure 3-7** Poincaré map of chaotic solutions to the forced two-well potential oscillator (3-2.10); 15,000 points.

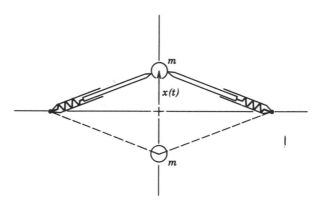

Figure 3-8 Sketch of a one-hinged arch. Forced snap-through oscillations can lead to chaotic vibrations (Clemens and Wauer, 1981).

In a similar example, Clemens and Wauer (1981) have analyzed the snap through oscillation of a one-hinged arch (Figure 3-8). Their equation takes the form

$$m\ddot{y} + \gamma\dot{y} + 2k\left(1 - \frac{1}{(b^2 + y^2)^{1/2}}\right)y = f_0\sin \omega t \qquad (3\text{-}2.11)$$

When only cubic nonlinearities are retained, this equation assumes the form (3-2.10) for the two-well potential Duffing oscillator.

Chaotic motions of an elasto-plastic arch have been studied by Poddar et al. (1987).

Chaos in a Pendulum

The motion of a particle in both space periodic and time periodic force fields serves as a model for several problems in physics. These include the classical pendulum, a charged particle in a moving electric field, synchronous rotors, and Josephson junctions. For example, the equation for the nonlinear dynamics of a particle in a traveling electric force field takes the form (e.g., see Zaslavsky and Chirikov, 1972)

$$\ddot{x} + \delta\dot{x} + \alpha \sin x = g(kx - \omega t) \qquad (3\text{-}2.12)$$

where $g(\ )$ is a periodic function. The study of the forced pendulum problem has revealed complex dynamics and chaotic vibrations (see Hockett and Holmes, 1985; Gwinn and Westervelt, 1985):

$$\ddot{x} + \delta\dot{x} + \alpha \sin x = f\cos \omega t \qquad (3\text{-}2.13)$$

Parametric oscillation is a term used to describe vibration of a system with time periodic changes in one or more of the parameters of a system. For example, a simply supported beam with sub-buckling axial compression is modeled by a one-mode approximation which yields an equation of the form

$$\ddot{x} + \omega_0^2(1 + \beta \cos \Omega t)x = 0 \qquad (3\text{-}2.14)$$

This linear ordinary differential equation is the well-known Mathieu equation. It is known that for certain values of ω_0^2, β, and Ω the equation admits unstable oscillating solutions. When nonlinearities are added, these vibrations result in a limit cycle. A similar example is the pendulum with a vibrating pivot point (Figure 3-9). Chaotic vibrations for this problem have been studied numerically by Levin and Koch (1981) and McLaughlin

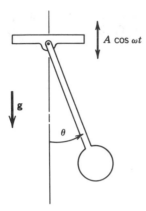

Figure 3-9   Parametrically forced pendulum.

(1981). The mathematical equation for this problem is

$$\ddot{\theta} + \beta\dot{\theta} + (1 + A\cos\Omega t)\sin\theta = 0 \qquad (3\text{-}2.15)$$

Period-doubling phenomena have been observed in numerical solutions and a Feigenbaum number was calculated for the sixth subharmonic bifurcation of $\delta = 4.74$.

Chaotic motion in a double pendulum have been studied by Richter and Scholz (1984).

Spherical Pendulum

The complex dynamics of a spherical pendulum with two degrees of freedom has been examined by Miles (1984a), who found chaotic solutions for this problem in numerical experiments when the suspension point undergoes forced periodic motions (Figure 3-9). The equations of motion can be derived from a Lagrangian given by

$$L = \tfrac{1}{2}m\left(\dot{x}^2 + \dot{y}^2 + \dot{z}^2\right) - m\frac{g}{l}(l - z) \qquad (3\text{-}2.16)$$

where l is the length of the pendulum and the coordinates (x, y, z) satisfy the constraint equation

$$(x - x_0)^2 + y^2 + z^2 = l^2$$

The suspension point is $x_0 = \epsilon l \cos\omega t$ and gravity acts in the z direction.

Miles uses a perturbation technique and transforms the resulting equation of motion using

$$x = \left[p_1(\tau)\cos\theta + q_1(\tau)\sin\theta \right] l\epsilon^{1/3}$$

$$y = \left[p_2(\tau)\cos\theta + q_2(\tau)\sin\theta \right] l\epsilon^{1/3}$$

(3-2.17)

where $\theta = \omega t$ and $\tau = \frac{1}{2}\epsilon^{2/3}\omega t$. The resulting set of four first-order equations for (p_1, p_2, q_1, q_2) with small damping added (represented by α) is found to be

$$\frac{d}{dt}\begin{bmatrix} p_1 \\ q_1 \\ p_2 \\ q_2 \end{bmatrix} = \begin{bmatrix} -\alpha & -\beta & -\delta & 0 \\ \beta & -\alpha & 0 & -\delta \\ \delta & 0 & -\alpha & -\beta \\ 0 & \delta & \beta & -\alpha \end{bmatrix}\begin{bmatrix} p_1 \\ q_1 \\ p_2 \\ q_2 \end{bmatrix} + \begin{bmatrix} 0 \\ 1 \\ 0 \\ 1 \end{bmatrix}$$

(3-2.18)

where α, β, and δ depend on the variables (p_1, q_1, p_2, q_2). The reader is referred to Miles (1984a) for the definitions of α, β, and δ. The divergence of this flow in the four-dimensional phase space is $\nabla \cdot \mathbf{f} = -4\alpha$. Equilibrium points of the set of equations (3-2.18) correspond to either periodic planar or nonplanar motions. Numerical simulation of this set of equations show a transition from closed orbit trajectories and discrete spectra to complex orbits and broad spectra characteristic of chaotic motions (Figure 3-9).

The Kicked Rotor

As we have seen in the examples of the horseshoe map (Figure 1-21) or the logistic equation (1-3.6) in Chapter 1, the nature of the chaotic dynamics is best uncovered by taking a Poincaré section of a continuous time flow in phase space. However, for most differential equation models of physical systems, it is impossible to obtain analytical results. An exception to this is the class of problems with pulsed forces, torques, or voltages. In the example considered here, we imagine a rotor with rotary inertia J and damping c which is subject to both a steady torque $c\omega_0$ and a periodic series of *pulsed torques* (see also Schuster, 1984). The equation of motion representing the change in angular momentum of the rotor is given by

$$J\dot{\omega} + c\omega = c\omega_0 + T(\theta)\sum_{n=-\infty}^{+\infty}\delta(t - n\tau), \quad \text{with } \dot{\omega} = \theta \quad (3-2.19)$$

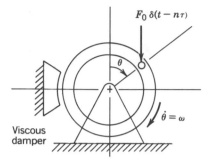

Figure 3-10 Rotor with viscous damping and periodically excited torque studied by Zaslavsky (1978).

The term $\delta(t - n\tau)$ represents the classical delta function which is zero everywhere except at $t = n\tau$ and whose area is unity. Thus, for times $n\tau - \epsilon < t < n\tau + \epsilon$, where $\epsilon \ll 1$, the angular momentum change is given by

$$J(\omega^+ - \omega^-) = T(\theta(n\tau)) \qquad (3\text{-}2.20)$$

For example, if the torque is created by a vertical force as shown in Figure 3-10, then the pulsed torque is proportional for $T(\theta) = F_0 \sin \theta$.

When $T(\theta) = 0$, Eq. (3-2.19) has a steady solution $\omega = \omega_0$, $\theta = \omega_0 t$.

To obtain a Poincaré map, we take a section right before each pulsed torque. Thus, we define $\theta_n = \theta(t = n\tau - \epsilon)$, $\epsilon \to 0^+$. One can relate (θ_n, ω_n) to $(\theta_{n+1}, \omega_{n+1})$ by solving the linear differential equation between pulses and using the jump in angular momentum condition (3-2.20) across the pulse. Between pulses, the rotation rate has the following behavior:

$$\omega = \omega_0 + ae^{-ct/J}$$

Carrying out this procedure, one can derive the following exact Poincaré map for the system (3-2.19):

$$\omega_{n+1} = \frac{c\tau}{J}\omega_0 + \omega_n - \frac{c}{J}(\theta_{n+1} - \theta_n) + \frac{1}{J}T(\theta_n)$$

$$\qquad (3\text{-}2.21)$$

$$\theta_{n+1} = \omega_0\tau + \theta_n + \frac{J}{c}(1 - e^{-c\tau/J})\left(\omega_n + \frac{1}{J}T(\theta_n) - \omega_0\right)$$

These equations were first derived by the Soviet physicist Zaslavsky (1978) to treat the nonlinear interaction between two oscillators. In this mechanical analog of his problem, ω_0 represents the frequency of one uncoupled oscillator (see also Ott, 1981, for a derivation).

This two-dimensional map is often nondimensionalized using

$$x_n = \frac{\theta_n}{2\pi} \quad (\text{mod } 1),$$

$$y_n = \frac{\omega_n - \omega_0}{\omega_0}$$

For $T(\theta) = F_0 \sin \theta$, and $\epsilon = F_0/J\omega_0$ Eqs. (3-2.21) then become

$$y_{n+1} = e^{-\Gamma}(y_n + \epsilon \sin 2\pi x_n) \tag{3-2.22}$$

$$x_{n+1} = \left\{ x_n + \frac{\Omega}{2\pi} + \frac{\Omega}{2\pi\Gamma}(1 - e^{-\Gamma})y_n + \frac{K}{\Gamma}(1 - e^{-\Gamma})\sin 2\pi x_n \right\}$$

where the braces { } indicate that only the fractional part is used (i.e., mod 1 or $0 \le \theta < 2\pi$). Also, $K = \epsilon\Omega/2\pi$, $\Gamma = c\tau/J$, and $\Omega = \omega_0\tau$. Here y_n measures the departure of the speed from the unperturbed equilibrium speed $\omega = \omega_0$. Note that this map contracts areas for $\Gamma > 0$ and preserves areas for $\Gamma = 0$.

This system of two difference equations has been found to exhibit chaotic solutions only if the following conditions are satisfied when ϵ is small:

$$1 < \frac{\Gamma}{1 - e^{-\Gamma}} < K \tag{3-2.23}$$

A typical case is shown in Figure 3-11 for the parameters $\Gamma = 5$, $\epsilon = 0.3$, $\Omega = 100$, and $K = 9$.

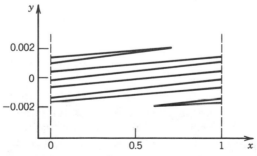

Figure 3-11 Strange attractor for the Zaslavsky map (3-2.22) for the kicked rotor in Figure 3-10: x represents normalized angular rotation and y represents the angular speed.

The problem of a kicked or pulsed double rotor with two degrees of freedom has been investigated by Kostelich et al. (1985, 1987).

Circle Map

A simpler version of the Zaslavsky map for two coupled oscillators can be obtained by letting the damping become larger, $\Gamma \gg 1$. In this limit, one can ignore the changes in ω or y (note that Δy is small in Figure 3-11). This leads to a one-dimensional map known as a *circle map*:

$$x_{n+1} = \left\{ x_n + \frac{\Omega}{2\pi} + \frac{K}{\Gamma}\sin 2\pi x_n \right\} \tag{3-2.24}$$

This equation has received extensive study (e.g., see Rand et al., 1982).

Other Rigid Body Problems

Three dimensional chaotic motions of a rigid body using Euler's equations (4-2.4) have been studied by Leipnik and Newton (1981), Szczygielski and Schweitzer (1985) and Wisdom et al. (1984). In the latter example, chaotic rotations of Hyperion, a moon of Jupiter, were predicted.

Another class of rigid body problems are those involving ship dynamics under wave excitation, (Nayfeh and Khdeir (1986), Virgin (1987)).

Aeroelastic Flutter

An example of chaos in autonomous mechanical systems is the flutter resulting from fluid flow over an elastic plate. This problem is known as *panel flutter* and readers are referred to a book by Dowell (1975) for more discussion of the mechanics of this problem. Panel flutter occurred on the outer skin of the early Saturn rocket flights that put men on the Moon in the early 1970s. Dowell and coworkers have done extensive numerical simulation of panel flutter. In earlier work, Kobayashi (1962) and Fung (1958) had observed nonperiodic motions in their analyses. In one set of problems, they looked at the combined effects of in-plane compression in the plate and fluid flow. More recent numerical results are given in Figure 3-12, showing stable phase plane trajectories for one set of fluid velocity and compressive load conditions and chaotic vibrations for another set of conditions (see also Dowell, 1982). This example also illustrates a different type of Poincaré map. Since there is no intrinsic time, one must choose a hyperplane in phase space and look at points where the trajectory penetrates

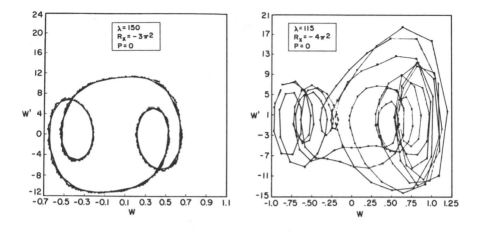

Figure 3-12 Flow over a buckled elastic plate. *Left*: Periodic aeroelastic vibrations. *Right*: Chaotic vibrations of the plate (from Dowell, 1982).

that plane. Dowell has done this for the panel flutter problem and has shown strange attractor-type Poincaré maps.

Nonlinear Electrical Circuits

One of the first examples of chaos in an electrical circuit to be discovered is that of a nonlinear inductor in a circuit treated by Ueda (1979). The equation for a circuit with nonlinear inductance and linear resistor, driven by a harmonic voltage, is found to be

$$\ddot{x} + k\dot{x} + x^3 = B \cos t \qquad (3\text{-}2.25)$$

which is another special case of Duffing's equation. Ueda of Kyoto University in Japan has obtained beautiful Poincaré maps of the chaotic dynamics of this equation using analog and digital simulation (Figure 3-13).

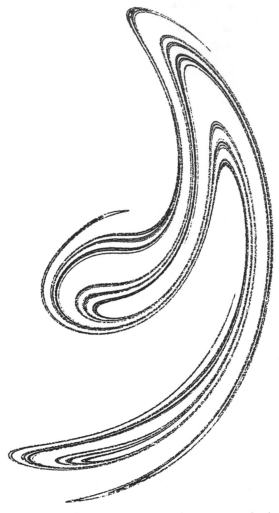

Figure 3-13   Poincaré map of chaotic oscillations (analog computer simulation) of a circuit with a nonlinear inductor [From Ueda (1979)].

Ueda has also modeled a negative resistor oscillator, shown in Figure 3-14. The equation for this system is a modified Van der Pol equation:

$$\ddot{x} + (x^2 - 1)\dot{x} + x^3 = B \cos \omega t \qquad (3\text{-}2.26)$$

It is interesting to note that both the Duffing and Van der Pol equations have been studied for decades yet nowhere in any of the standard references on nonlinear vibrations are chaotic solutions reported. Other nonlinear chaotic circuits are discussed in the next section.

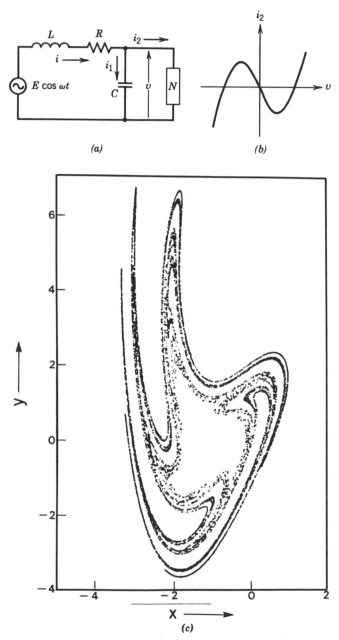

Figure 3-14 Poincaré map of chaotic analog computer simulation of a forced Van der Pol type circuit [from Ueda and Akamatsu (1981)].

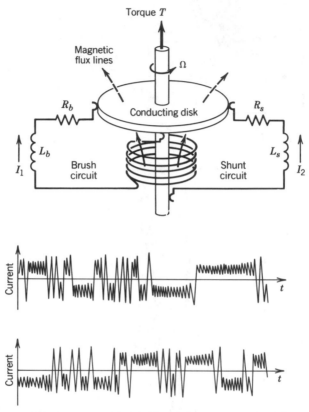

Figure 3-15 *Top*: Disk dynamo model of Robbins (1977) for reversals of the earth's magnetic field. *Bottom*: Chaotic current reversals from numerical solutions of disk dynamo equations (3-2.27).

Magnetomechanical Models

A physical model that has received considerable attention is the rotating disk in a magnetic field. This system is of interest to geophysicists as a potential model to explain reversals of the earth's magnetic field. A single-disk dynamo is shown in Figure 3-15. The equations governing the rotation Ω and the currents I_1 and I_2 are of the form (see Robbins, 1977)

$$J\dot{\Omega} = -k\Omega - \mu_2 I_1(I_1 + I_2) + T$$

$$L_1 \dot{I}_1 = -RI_1 - R_3 I_2 + \mu_1 \Omega I_1 \qquad (3\text{-}2.27)$$

$$L_2 \dot{I}_2 = -R_2 I_2 + \mu_2 \Omega I_1$$

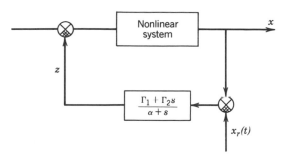

Figure 3-16 Feedback control system: nonlinear plant with linear feedback control.

where T is an applied constant torque. The time traces in Figure 3-15 show that the current (and hence the magnetic field) can reverse in an apparently random manner.

Control System Chaos

Imagine a mechanical device with a nonlinear restoring force and suppose a control force is added to move the system from one position to another according to some prescribed reference signal $x_r(t)$. Such a system can be modeled by the following third-order system:

$$m\ddot{x} + \delta\dot{x} + F(x) = -z$$

$$\dot{z} + \alpha z = G_1[x - x_r(t)] + G_2\dot{x}$$

(3-2.28)

Here z represents a feedback force, and G_1 and G_2 represent position and velocity feedback gains, respectively. This system of equations can be represented by the block diagram in Figure 3-16 with a nonlinear mechanical plant and a linear feedback law.

Two types of chaotic vibrations problem can be explored here. First, if the system is autonomous, that is, the reference signal is zero—$x_r(t) = 0$, one could explore the gain space (G_1, G_2) for regions of steady, periodic, and chaotic vibrations. The second problem arises if $x_r(t)$ is periodic. That is, we wish to move the mass through a given path over and over again as in some manufacturing robotic device. One could then explore the parameters of frequency and gain for which the system is periodic or chaotic as in Figure 3-17.

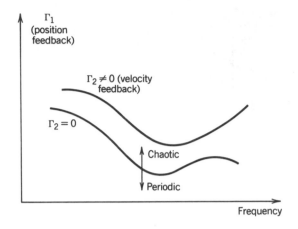

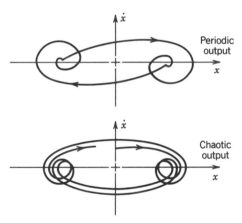

Figure 3-17 *Top*: Chaos boundary as a function of feedback gain and input command frequency. *Bottom*: Trajectories of periodic and chaotic dynamics for a mass with feedback control and nonlinear restoring force with a dead-band region (see Figure 3-4).

Chaotic vibrations for an autonomous system of the form (3-2.28) were studied by Holmes and Moon (1983) as well as by Holmes (1984). For example, when $F(x) = x(x^2 - 1)(x^2 - B)$, the mechanical system has three stable equilibria. This system has been shown to exhibit both periodic limit cycle oscillation as well as chaotic motion.

The problem of a forced feedback system has been studied by Golnaraghi and Moon (1986). Also Sparrow (1981) looked into chaotic oscillations in a system with a piecewise linear feedback function. Other examples have since appeared in the literature e.g. Baillieul et al. (1980), Brockett (1982).

3.3   PHYSICAL EXPERIMENTS IN CHAOTIC SYSTEMS

Early Observations of Chaotic Vibrations

Early scholars in the fields of electrical and mechanical vibrations rarely mention nonperiodic, sustained oscillations with the exception of problems relating to fluid turbulence. Yet chaotic motions have always existed. Experimentalists, however, were not trained to recognize them. Inspired by theoreticians, the engineer or scientist was taught to look for resonances and periodic vibrations in physical experiments and to label all other motions as "noise."

Joe Keller, a mathematician at Stanford University, has speculated on the reason for the apparent myopic vision of experimental scientists as regards chaotic phenomena in the last century. He notes that the completeness and beauty of linear differential equations led to its domination of the mathematical training of most scientists and engineers.

Examples of nonperiodic oscillations can be found in the literature, however. Three cases are cited here. First, Van der Pol and Van der Mark (1927) in a paper on oscillations of a vacuum tube circuit make the following remark at the end of their paper: "Often an irregular noise is heard in the telephone receiver before the frequency jumps." No explanation is offered for these events and in classical treatises on the Van der Pol oscillator, no further mention is made of "irregular noises."

Another observation of nonsteady vibrations was reported by Evensen (1967) in a NASA publication on shell vibrations. While the focus of this study was originally nonlinear oscillations of elastic cylindrical shells, in both analog computer and experiments nonsteady vibrations were observed for sufficiently large driving excitation. It is likely that this observation was not unique and that many other examples of chaos could be found in the literature where one or two lines of description of nonperiodic motions are buried in a paper on periodic vibrations.

In a more recent paper, Tseng and Dugundji (1971) studied the nonlinear vibrations of a buckled beam. The beam was rigidly clamped at both ends and then compressed to buckling. This created an arched structure. When the beam was vibrated transverse to its length and the acceleration forces increased, snap through occurred. In this regime, intermittent oscilla-

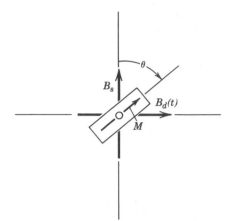

Figure 3-18 Sketch of a magnetic dipole rotor in crossed static and dynamic magnetic fields—a "magnetic pendulum."

tions were observed as well as subharmonic responses. The analysis in the paper, however, only dealt with periodic vibrations.

Many readers may recall similar phenomena in experiments that they have done or in engineering practice. Chaotic noise has always been around in engineering devices, but until recently we had no models or mathematics to simulate or describe it.

Rigid-Body Systems

The pendulum is such a classical paradigm in dynamics that one should be curious to find out if this paragon of deterministic dynamics can exhibit chaotic oscillations. To answer this question, the author and coworkers at Cornell University (Moon et al., 1987) constructed a magnetic dipole rotor with a restoring torque proportional to the sine of the angle between the dipole axis and a fixed magnetic field (Figure 3-18). A time-periodic restoring torque was provided by placing a sinusoidal voltage across two poles transverse to the steady magnetic field. The mathematical model for this forced magnetic pendulum becomes

$$J\ddot{\theta} + c\dot{\theta} + MB_s\sin\theta = MB_d\cos\theta \cos\Omega t \qquad (3\text{-}3.1)$$

where J is the rotational inertia of the rotor, c is a viscous damping constant, M is the magnetic dipole strength of the rotor dipole, and B_s and B_d are the intensities of the steady and dynamic magnetic fields, respectively. Figure 3-19 shows a comparison of periodic and chaotic rotor speeds under periodic excitation. Additional discussion of this experiment may be found in Chapters 4 and 6. Chaos theory has also been used to excite

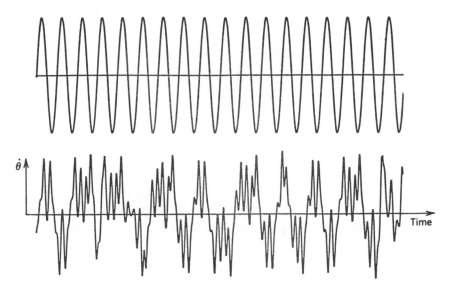

Figure 3-19 *Top*: Periodic motion of a magnetic rotor (Figure 3-18). *Bottom*: Chaotic motion of a magnetic rotor.

nonperiodic vibrations in a multi-pendulum mobile sculpture by Viet et al. (1983).

Magnetic Compass Needle

Another magnetomechanical device with stochastic dynamics is a compass needle in an oscillatory or rotating magnetic field (Figure 3-20). The rotating magnetic field can be created by two Helmholtz coils with sinusoidal currents applied with different phases to each coil. Croquette and Poitou (1981) have performed experiments on this problem. They have modeled this problem with the following equation and observed period-doubling bifurcations.

$$J\ddot{\theta} = -\mu[\sin(\theta - \omega t) + \sin(\theta + \omega t)] - \gamma\dot{\theta} \qquad (3\text{-}3.2)$$

Here damping is very small ($\gamma/J\omega \approx 10^{-2}$). This system is an example of a problem with near zero dissipation ($\gamma = 0$) and is referred to as a *Hamiltonian system*. As discussed in the example of the Fermi accelerator model, chaos in systems without dissipation is often referred to as *stochasticity*. Poincaré maps of these systems fill up regions of phase space in contrast to the Cantor set structure seen in dissipative systems with strange attractors.

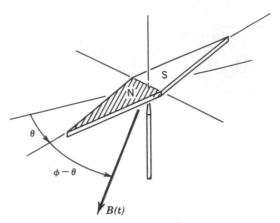

Figure 3-20 Magnetic dipole rotor in a rotating magnetic field.

Magnetically Levitated Vehicles

Suspension systems for land-based vehicles must provide vertical and lateral restoring forces when the vehicle departs from its straight path. Conventional suspension systems such as pneumatic tires and steel wheels on steel rails, as well as the futuristic systems of air cushion or magnetic levitation, all exhibit nonlinear stiffness and damping behavior and are thus candidates for chaotic vibrations. As an illustration, some experiments performed at Cornell University on a magnetically levitated vehicle are described. [See the book by Moon (1984a) which describes magnetic levitation transportation mechanics.]

In this experiment permanent magnets were attached to a rigid platform and a continuous L-shaped aluminum guideway was moved past the model using a 1.2 meter diameter rotating wheel (Figure 3-21). The induced eddy currents in the aluminum guideway interact with the magnetic field of the magnets to produce lift, drag, and lateral guidance forces. The magnetic drag force is nonconservative and can pump energy into the vibrations of the model. Thus, under certain conditions, the model can undergo limit cycle oscillations. As the speed is increased, damped vibrations change to growing oscillations (see bottom of Figure 3-21). The nonlinearities in the suspension forces limit the vibration and a limit cycle motion results. [This bifurcation in stability is known in mathematics as a *Hopf bifurcation* (Chapter 1). In mechanics it is called a *flutter* oscillation.]

In addition to flutter or limit cycle oscillations, the levitated model can undergo static bifurcations. Thus, at certain speeds, the equilibrium state can change from vertical to two stable tilted positions as shown in Figure 3-21. This latter instability is known in aircraft dynamics as *divergence* and

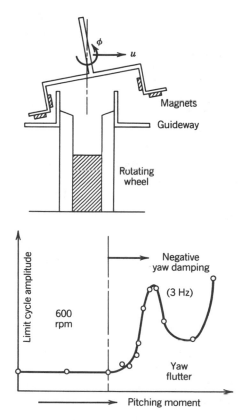

Figure 3-21 *Top*: Sketch of magnetically levitated model on a rotating aluminum guideway. *Bottom*: Limit cycle bifurcation of levitated model.

is analogous to buckling of an elastic column. In our experiments, chaotic vibrations occurred when the system exhibited both divergence (multiple equilibrium states) and flutter. The flutter provides a mechanism to throw the model from one side of the guideway to another, similar to the buckled beam problem discussed in Chapter 2. The mathematical model for this instability, however, has two degrees of freedom. Lateral and roll dynamics

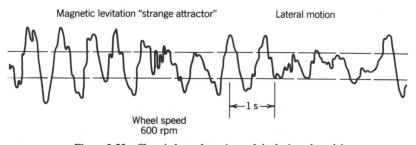

Figure 3-22 Chaotic lateral motions of the levitated model.

were measured from films of the chaotic vibrations (Figure 3-22)). These vibrations were quite violent and if they occurred in an actual vehicle traveling at 400–500 km/h, the vehicle would probably derail and be destroyed.

Chaos in Elastic Continua

Many experiments on chaotic vibrations in elastic beams have been carried out by the author and coworkers (e.g., see Moon and Holmes, 1979, 1985; Moon 1980a, b, 1984; Moon and Shaw, 1983). Two types of problems have been investigated. In the first problem, the partial differential equation of motion for the beam is essentially linear, but the body forces or boundary conditions are nonlinear. In the second problem, the motions are sufficiently large enough that significant nonlinear terms enter the equations of motion.

The equation of motion for an elastic beam with small slopes and deflections is given by

$$D\frac{\partial^4 v}{\partial x^4} + m\frac{\partial^2 v}{\partial t^2} = f\left(v, \frac{dv}{dt}, t\right) \tag{3-3.3}$$

where v is the transverse displacement of the beam, D represents an elastic stiffness, and m is the mass per unit length. The right-hand term represents the effects of distributed body forces or internal damping. In many of the experiments at Cornell University, we used permanent magnets to create nonlinear body force terms.

When the displacement and slope of the beam centerline are large, we use variables (u, v, θ) to characterize the horizontal and vertical displacements and the slope which are related by (see Figure 3-23))

$$(1 + u')^2 + (v')^2 = 1, \qquad \tan\theta = \frac{v'}{1 + u'} \tag{3-3.4}$$

where $(\ )' = \partial/\partial s$ and s is the length along the deformed beam. The balance of momentum equations then take the form

$$m\ddot{v} = f_v - G'$$
$$m\ddot{u} = f_u + H' \tag{3-3.5}$$

where

$$G = D\theta''(1 + u') - Tv'$$
$$H = D\theta''v' + T(1 + u')$$

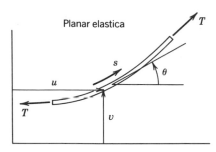

Figure 3-23 Planar deformation of an elastic rod.

In these equations, (f_u, f_v) represent body force components, while T represents the axial force in the rod. The nonlinearities in these equations are distinguished from those in fluid mechanics by the fact that no convective or kinematic nonlinearities enter the problem. Also, the local stress–strain relations are linear. The nonlinear terms arise from the change in geometric shape and are known as *geometric nonlinearities*. [See Love (1922) for a discussion of nonlinear rod theory.]

Magnetoelastic Buckled Beam. In this example, an elastic cantilevered beam is buckled by placing magnets near the free end of the beam (see Chapters 2 and 4 as well as Moon and Holmes, 1979; Moon, 1980a, b, 1984b). The magnetic forces destabilize the straight unbent position and create multiple equilibrium positions as shown in Figure 3-24a. In experiments, we have created up to four stable equilibrium positions with four magnets. In the postbuckled state, the system represents a particle in a two- or more well potential (Figure 1-2b). The whole system is placed on a vibration shaker and oscillates with constant amplitude and frequency. For small oscillations, the beam vibrations occur about one of the equilibrium

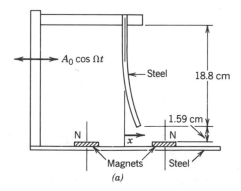

Figure 3-24 (a) Steel elastic beam on a periodically moving support that is buckled by magnetic body forces.

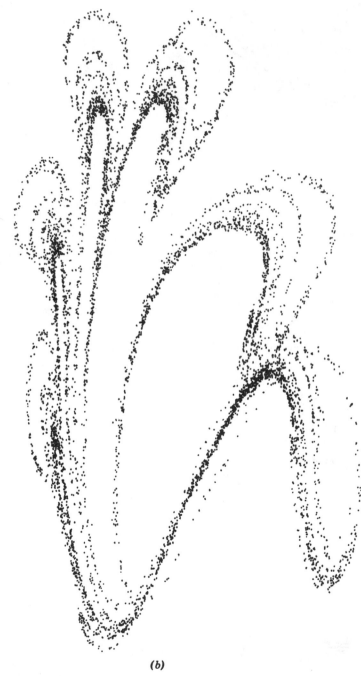

(b)

Figure 3-24 (b) Experimental Poincaré map of chaotic motion of the magnetically buckled beam, Fleur de Poincaré.

positions. As the amplitude is increased, however, the beam can jump out of the potential well and chaotic motions can occur, with the beam jumping from one well to another (Figure 3-6). A Poincaré map of this phenomenon is shown in Figure 3-24b. (We call this map the Fleur de Poincaré.)

The equation used to model this system is a modal approximation to the beam equation (3-3.3) with nonlinear magnetic forces acting at the tip.

A one-mode approximation for a damped beam with a free end gives good results. This equation can be rewritten as three first-order equations. Note that here the x variable refers to nondimensional modal amplitude and *not* to the distance along the beam.

$$\dot{x} = y$$

$$\dot{y} = -\gamma\dot{x} + \tfrac{1}{2}x(1 - x^2) - A_0\omega^2\cos z \qquad (3\text{-}3.6)$$

$$\dot{z} = \omega$$

This problem is analogous to a particle in a double-well potential $\mathscr{V} = -(x^2 - x^4/2)/4$. This experiment is discussed throughout this book. The Poincaré section (Figure 3-24b) has the character of two-dimensional point mappings. Typically, the experiments *did not* exhibit period doubling before the motion became chaotic. Odd subharmonics were often a precursor to chaos.

Another variation of this experiment is an inverted pendulum with an elastic spring reported in the People's Republic of China by Zhu (1983) from Beijing University. For a weak spring, the inverted pendulum has two stable equilibria similar to the two-well potential problem.

Two-Degree-of-Freedom Buckled Beam.   To explore the effects of added degrees of freedom, we built an elastic version of the spherical pendulum (Figure 3-9) where a beam with circular cross section was used (see Moon, 1980b). Again magnets were used to buckle the beam but the tip was free to move in two directions. This introduced two incommensurate natural frequencies and quasiperiodic vibrations occurred which eventually became chaotic (Figure 3-25).

This experimental system can be modeled by equations for two coupled oscillators as given by

$$\ddot{x} + \gamma\dot{x} - \tfrac{1}{2}x(1 - x^2) + \beta xy^2 = f_2 \qquad (3\text{-}3.7a)$$

$$\ddot{y} + \delta\dot{y} + \alpha(1 + \epsilon y^2)y + \beta x^2 y = f_0 + f_1\cos \omega t \qquad (3\text{-}3.7b)$$

The terms f_0 and f_2 account for gravity if the beam is not initially parallel with the earth's gravitational field, and the coupling terms are conservative.

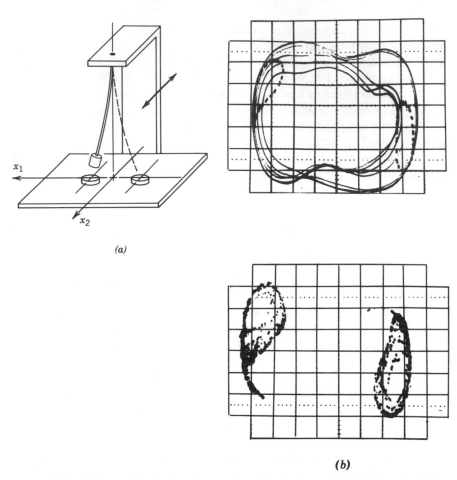

(a)

(b)

Figure Figure 3-25 (*a*) Sketch of an elastic rod undergoing three-dimensional motions in the neighborhood of a double-well potential created by two magnets. (*b*) *Top*: Simultaneous time trace of phase plane motion and Poincaré map of quasiperiodic motion. *Bottom*: Poincaré map of chaotic motion.

If the coupling is small, one can solve for $y(t)$ from (3-3.7b) and the equation for $x(t)$ looks like a parametric oscillator.

Miles (1984b) has performed numerical experiments on two quadratically coupled, damped oscillators and has found regions of chaotic motions resulting from sinusoidal forcing. He examined the special case when the two linear natural frequencies ω_1 and ω_2 were related by $\omega_2 \simeq 2\omega$.

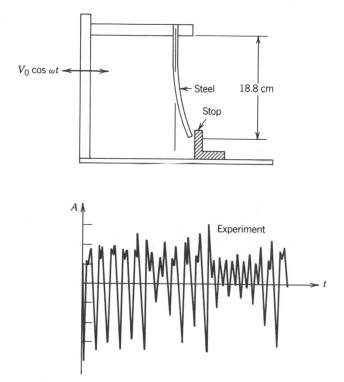

Figure 3-26   Chaotic vibrations of an elastic beam with a nonlinear boundary condition.

Elastic Beam with Nonlinear Boundary Conditions.   Multiple equilibrium positions are not needed in a mechanical system to get chaotic vibrations. Any strong nonlinearity will likely produce chaotic noise with periodic inputs. One example of a system with one equilibrium position is an elastic beam with nonlinear boundary conditions (see Moon and Shaw, 1983). Nonlinear boundary conditions are those that depend on the motion. For example, suppose the end is free for one direction of motion and is pinned for the other direction of motion. The chaotic time history of this beam is shown in Figure 3-26. Another variation of this problem is a two-sided constraint with play which gives three different linear regimes for the bending of the beam. Experiments in our laboratory also show chaos for this nonlinear boundary condition. Shaw (1985a, b) has performed an analysis of these mechanical oscillations when play or a dead zone is present.

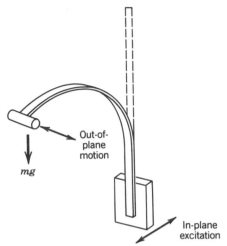

Figure 3-27 Sketch of elastica showing in-plane and out-of-plane motions.

Three-Dimensional Elastica and Strings

Under certain conditions, the forced planar motion of the nonlinear elastica described by (3-3.5) becomes unstable and three-dimensional motions result. Similar phenomena are known for the planar motion of a stretched string (Miles, 1984b). At Cornell University, we have performed several experiments with very thin flexible steel elastica with rectangular cross section (e.g., 0.25 mm × 10 mm × 20 cm long) known as "Feeler" gauge steel strips (Figure 3-27). For these beams, small motion in the stiff or lateral direction of the unbent beam is nearly impossible without buckling or twisting of the local cross sections. However, when there is significant bending in the weak direction, lateral displacements are possible accompanied by twisting of the local cross sections. We have shown that planar vibrations of the beam in the weak direction near one of the many natural frequencies not only become unstable but can exhibit chaotic motions as well. This is demonstrated in Figure 3-28 where power spectra (FFT, see Chapter 4) show a broad spectrum of frequencies when the driving input has a single frequency input. Similar phenomena are observed for very thin sheets of paper. In fact, we have shown that chaotic motions of very thin sheets of paper generate a broad spectrum of acoustic noise in the surrounding air.

Impact Print Hammer

Impact-type problems have emerged as an obvious class of mechanical examples of chaotic vibrations. The bouncing ball (3-2.9), the Fermi accel-

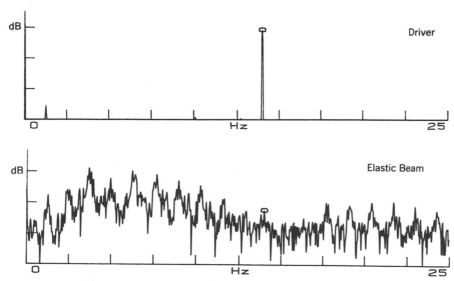

Figure 3-28 Fourier spectra for forced vibrations of a thin elastic beam. Broad spectrum chaos is the result of out-of-plane vibration (Moon and Cusumano, 1987).

erator model (3-2.8), and the beam with nonlinear boundary conditions all fall into this category. A practical realization of impact-induced chaotic vibrations is the impact print hammer experiment studied by Hendriks (1983) (Figure 3-29). In this printing device, a hammer head is accelerated by a magnetic force and the kinetic energy is absorbed in pushing ink from a ribbon onto paper. Hendriks uses an empirical law for the impact force versus relative displacement after impact; u is equal to the ratio of

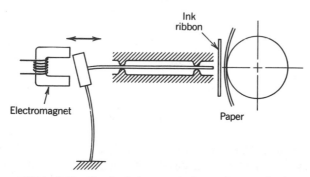

Figure 3-29 Sketch of pin-actuator for a printer mechanism.

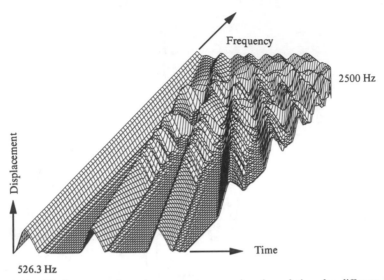

Figure 3-30 Displacement of a printer actuator as a function of time for different input frequencies showing loss of predictable output [from Hendriks (1983); Copyright 1983 by International Business Machines Corporation; reprinted with permission].

displacement to ribbon–paper thickness:

$$F = -AE_p u^{2.7}, \qquad \dot{u} > 0$$

$$= -AE_p \beta u^{11}, \qquad \dot{u} < 0 \qquad (3\text{-}3.8)$$

where A is the area of hammer–ribbon contact, E_P acts like a ribbon–paper stiffness, and β is a constant that depends on the maximum displacement. The point to be made is that this force is extremely nonlinear.

When the print hammer is excited by a periodic voltage, it will respond periodically as long as the frequency is low. But as the frequency is increased, the hammer has little time to damp or settle out and the impact history becomes chaotic (see Figure 3-30). Thus, chaotic vibrations restrict the speed at which the printer can work. One potential solution which is under study is the addition of feedback control to suppress this chaos.

Nonlinear Circuits

Periodically Excited Circuits: Chaos in a Diode Circuit. The idealized diode is a circuit element that either conducts or does not. Such on–off behavior represents a strong nonlinearity. A number of experiments in

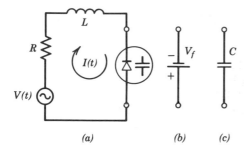

Figure 3-31 (*a*) Model for a varactor diode circuit. (*b*) Circuit element when the diode is conducting. (*c*) Circuit element when the diode is off [from Rollins and Hunt (1982) with permission of The American Physical Society, copyright 1982].

chaotic oscillations have been performed using a particular diode element called a varactor diode (Lindsay, 1981; Testa et al., 1982; Rollins and Hunt, 1982) using a circuit similar to the one in Figure 3-31. Both period doubling and chaotic behavior were reported. The period doubling suggests that an underlying mathematical model is a one-dimensional map in which the absolute value of the maximum current value in the circuit during the $(n + 1)$st cycle depends on that in the nth cycle:

$$|I_{max}|_{n+1} = F(|I_{max}|_n) \qquad (3\text{-}3.9)$$

One of the interesting questions regarding this system was the physical origin of the nonlinearity. In the earlier work of Lindsay, it was proposed that the diode could be modeled as a highly nonlinear capacitance, where

$$c = c_0(1 - \alpha V)^{-\gamma}$$

$$\frac{d}{dt}c(V)V = I \qquad (3\text{-}3.10)$$

$$L\frac{dI}{dt} = -RI - V + V_0\sin \omega t$$

where $\gamma \simeq 0.44$. Rollins and Hunt (1982), however, have proposed an entirely different model in which the circuit acts as either one of two linear circuits, shown in Figure 3-31*b*, *c*. Each cycle consists of a conducting and a nonconducting phase. The nonlinearity arises in determining when to switch from the conducting circuit with bias voltage V_f to the nonconducting circuit with constant capacitance. The switching time is a function of the maximum current value $|I_{max}|$. In this model, exact solutions of the circuit differential equations are known in each interval, with unknown constants to be determined using continuity of current and voltage at the switching times. Rollins and Hunt use this technique to calculate numeri-

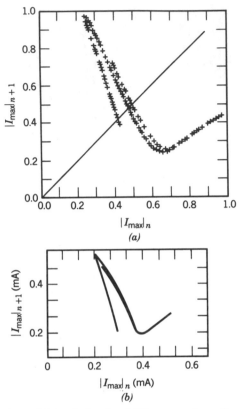

Figure 3-32 Comparison of (a) calculated and (b) measured one-dimensional maps for the varactor diode circuit of Figure 3-31 [from Rollins and Hunt (1982) with permission of The American Physical Society, copyright 1982].

cally the mapping function shown in Figure 3-32. Later experiments showed that this model accounted for more of the physics than the earlier version using nonlinear capacitance.

This example illustrates a problem in nonlinear dynamics. In the rush to explain chaotic dynamics in physical systems, there is a temptation to propose mathematical models that emulate the classic chaos paradigms more than the actual physics of the system. This could be forgiven in the early days of discovery and exploration in the subject. But, as the field of nonlinear dynamics matures, more accountability for both the mathematical and *physical* principles underlying the phenomena must be required. The connection between physical laws (e.g., Newton's laws and Maxwell's equations) and mathematical models should eventually become transparent if the new explanations of chaotic phenomena are to be accepted.

The nonlinear circuit with a varactor diode was discussed in the previous section and several experimental papers reporting chaos in this system were cited. Here we report on one particular experiment by Bucko et al. (1984) who looked at a series circuit with a diode, inductor, and resistor driven by a sinusoidal voltage which has a mathematical model of the form

$$L\frac{dI}{dt} + RI + f\left(I, \int I dt\right) = V_0 \cos \omega t \qquad (3\text{-}3.11)$$

where the properties of the nonlinear diode $f(I)$ were discussed in the previous section. Bucko et al. explored the parameter plane (v_0, ω) and outlined regions of subharmonic and chaotic response. These results are shown in Figure 3-33. Figure 3-33a shows a driving frequency range $0.5 < \omega/2\pi < 4.0$ MHz. These data show that one can choose a parameter path that results in a period-doubling route to chaos. However, one can also follow paths that apparently do not follow this route. Figure 3-33a also shows chaotic islands which when expanded in Figure 3-33b exhibit further islands of chaos. This example shows that when the basic equations (3-3.11) are three differential equations, the Poincaré map of the dynamics is *two* dimensional and the period-doubling properties of the one-dimensional map may not hold in such systems.

For certain parameter regimes, however, the two-dimensional map may look one dimensional and the dynamics are likely to behave as a one-dimensional noninvertible map. The experimental moral of this is the following: When there is more than one essential nondimensional group in a physical problem, one should explore a region of parameter space to uncover the full range of possibilities in the nonlinear dynamics.

Nonlinear Inductor.   Bryant and Jeffries (1984a) have studied a sinusoidally driven circuit with a linear negative resistor and a nonlinear inductor with hysteresis. In this work, they looked at four circuit elements in parallel: a voltage generator, negative resistor, capacitor, and a coil around a toroidal magnetic core, with typical values of $C \approx 7.5$ µF, $R = -500$ Ω, and a forcing frequency of around 200 Hz or higher. The negative resistor was created by an operational amplifier circuit. If N is the number of turns around the inductor, A the effective core cross section, and l the magnetic path length, the equation for the flux density B in the core is given by

$$NAC\ddot{B} + \frac{NA}{R}\dot{B} + \frac{l}{N}H(B) = I(t) \qquad (3\text{-}3.12)$$

where $H(B)$ is the nonlinear magnetic field constitutive relation of the core material. In their experiments, they used $N = 100$ turns, $A \approx 1.5 \times 10^{-5}$ m^2, and $l \approx 0.1$ m.

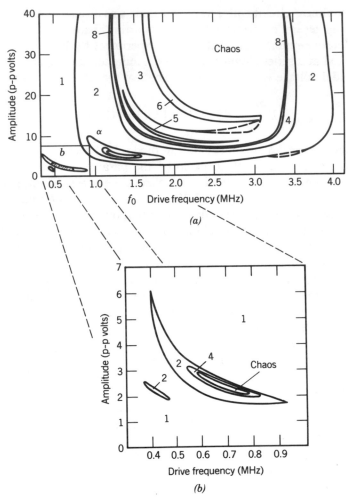

Figure 3-33 (*a*) Subharmonic and chaotic oscillation regions in the driver voltage amplitude–frequency plane for an inductor–resistor–diode series circuit. (*b*) Enlargement of diagram in (*a*) [from Bucko et al. (1984) with permission of Elsevier Science Publishers, copyright 1984].

Using this circuit, they observe quasiperiodic vibrations, phase-locked motions, period doubling, and chaotic oscillations.

Autonomous Nonlinear Circuits. Autonomous chaotic oscillations in a tunnel diode circuit have been observed by Gollub et al. (1980) for the circuit shown in Figure 3-34*a*.

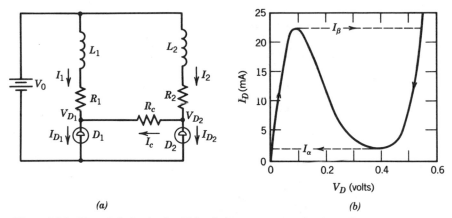

(a)   *(b)*

Figure 3-34 Tunnel diode circuit which admits autonomous chaotic oscillations [from Gollub et al. (1980) with permission of Plenum Publishing Corp., copyright 1980].

The nonlinear elements in this circuit are two tunnel diodes. The current–voltage relation shown in Figure 3-34*b* is obviously nonlinear and exhibits a hysteresis loop for cyclic variations in current I_D. In this work, the authors use return maps to construct pseudo-phase-plane Poincaré maps. That is, they time sample the current

$$x_n \equiv I_{D_2}(t_0 + n\tau) \qquad (3\text{-}3.13)$$

where n is an integer, and then plot x_n versus x_{n+1}. The data were sampled when the voltage V_{D_1} passed through a value of 0.42 V in the decreasing sense. The authors also use Fourier spectra and calculation of Lyapunov constants to measure the divergence rate of nearby trajectories.

As noted in the section on mathematical models, Ueda (1979) studied chaos in a circuit with negative resistance. A novel way to achieve negative resistance in the laboratory is with an operational amplifier. Two examples of experiments on chaotic oscillations in nonlinear circuits using this technique are those by Matsumoto et al. (1984, 1985) and Bryant and Jeffries (1984a, b).

The circuit studied by Matsumoto et al. is shown in Figure 3-35*a* and consists of three coupled current circuits with a nonlinear resistor. This circuit is autonomous; that is, there is no driving voltage. Thus, the system can produce oscillations only if the nonlinear resistor has negative resistance over some voltage range. In their model, Matsumoto et al. (1984) chose a trilinear current–voltage relation shown in Figure 3-35*b* which has

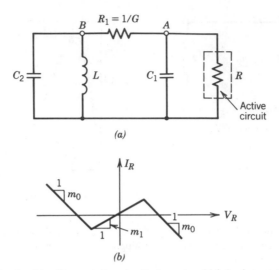

(a)

(b)

Figure 3-35 Circuit with trilinear active circuit elements which leads to autonomous chaotic oscillations [from Matsumoto et al. (1985) © 1984 Institute of Electrical and Electronic Engineers].

the form

$$g(V_1) = m_0 V_1 + \tfrac{1}{2}(m - m_0)|V_1 + b| + \tfrac{1}{2}(m_0 - m_1)|V_1 - b| \quad (3\text{-}3.14)$$

The resulting circuit equations are obtained by summing currents at nodes A and B in Figure 3-35a and voltages in the left-hand circuit loop.

$$C_1 \dot{V}_1 = \frac{1}{R}(V_2 - V_1) - g(V_1)$$

$$C_2 \dot{V}_2 = \frac{1}{R}(V_1 - V_2) - I \qquad (3\text{-}3.15)$$

$$L\dot{I} = -V_2$$

where V_1 and V_2 are the voltages across the capacitors C_1 and C_2 and I is the current through the inductor. Chua and coworkers created the trilinear resistor (3-3.14) by using an operational amplifier with diodes (see Chapter 4 for details). For small voltages, the nonlinear resistance is negative, and the equilibrium position $(V_1, V_2, I) = (0, 0, 0)$ is unstable and oscillations occur. Chaotic oscillations were found for $1/C_1 = 9$, $1/C_2 = 1$, $1/L = 7$,

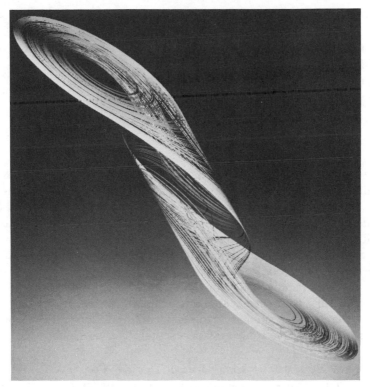

Figure 3-36 Chaotic trajectory for a circuit with a trilinear resistor (see Figure 3-35) numerical simulation. This attractor, based on Chua's circuit, is called the "double scroll" [from Matsumoto et al. (1985) © 1985 Institute of Electrical and Electronic Engineers].

$G = 0.7$, $m_0 = -0.5$, $m_1 = -0.8$, and $b = 1$ in a set of consistent units. A chaotic time history is shown in Figure 3-36, which has the same character as the Lorenz attractor (Figure 1-25).

Chaotic Dynamics in Fluid Systems

While the primary focus of this book is on low-order mechanical and electrical systems, the major impact of the new dynamics on fluid mechanics warrants mention of at least a few fluid experiments in chaotic motions. We recall from Chapter 1 that the major nonlinearity in fluid problems is a convective acceleration term $\mathbf{v} \cdot \nabla \mathbf{v}$ in the equations of motion (1-1.3). However, other nonlinearities may also play a role such as free surface or interface conditions and non-Newtonian viscous effects. We can classify five types of fluid experiments in which chaotic motions have been observed

1. Closed-flow systems: Rayleigh–Benard convection, Taylor–Couette flow between cylinders
2. Open-flow systems: pipe flow, boundary layers, jets
3. Fluid particles: dripping faucet
4. Waves on fluid surfaces: gravity waves
5. Reacting fluids: chemical stirred tank reactor

One reason for the intense interest in chaotic dynamics and fluids is its potential for unlocking the secrets of turbulence. [For example, see Swinney (1983) for a review and the edited volume by Tatsumi (1984) for a collection of papers on fluids and chaos.] Some feel that this may be too ambitious a goal for a theory based on a few ordinary differential equations and maps. One view is that dynamical systems theory will provide a good model for the transition to turbulence (called by some "weakling" turbulence) but will require major breakthroughs to solve the more difficult problem of fully developed spatial and temporal turbulence (strong turbulence). Whatever the ultimate progress, nonlinear dynamical theory has added new tools to the study of experimental fluid mechanics.

Closed-Flow Systems: Rayleigh–Benard Thermal Convection. We recall from Chapter 1 that a thermal gradient in a fluid under gravity produces a buoyancy force that leads to a vortex-type instability with resulting chaotic and turbulent motions. By far the most studied experimental system is the thermal convection of fluid in a closed box. This is the system that Lorenz tried to model with his famous equations (3-2.3).

Experimental studies of Rayleigh–Benard thermal convection in a box have shown period-doubling sequences as precursors to the chaotic state. They have been carried out in helium, water, and mercury for a wide range of nondimensional Prandtl numbers and Rayleigh numbers. These experiments emerged in the late 1970s. For example, Libchaber and Maurer (1978) observed period-doubling convection oscillation in helium. A number of experimental papers have emerged from a group at the French National Laboratory at Saclay, France, associated with Bergé and co-workers (1980, 1982, 1985) See also Dubois et al. (1982). The experiment is similar to that pictured in Figure 3-1 with a fluid of silicone oil in a rectangular cell with dimensions 2 cm × 2.4 cm × 4 cm. They have observed both the quasiperiodic route to chaos (Newhouse et al., 1978) and intermittent chaos. In the former, they observe the following sequence of dynamic events as the temperature gradient is increased:

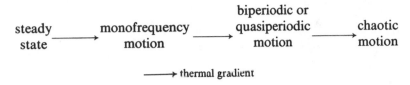

The frequency range observed in their experiments is very low, for example, $9-30 \times 10^{-3}$ Hz. They were one of the first groups to obtain Poincaré maps in fluid experiments. This was facilitated by their discovery of regions in the flow where one frequency or oscillator was predominant. Thus, they could use one frequency to synchronize the Poincaré maps. Two maps are shown in Figure 2-18. The first is quasiperiodic and the frequency ratio is close to 3. The second is based on 1500 Poincaré points and shows a breakup of the toroidal attractor before chaos sets in. The techniques used to measure the motion included laser Doppler anemometry and a differential interferometric method. More recent work involving mode locking and chaos in convection problems has been done by Haucke and Ecke (1987).

Taylor–Couette Flow Between Cylinders.   A classic fluid mechanics system which exhibits preturbulent chaos is the flow between two rotating cylinders (called Taylor–Couette flow) shown in Figure 3-37. Much work has been done on this system (e.g., see Swinney, 1983, for a review). This flow is sensitive to the Reynolds number $R = (b - a)a\Omega_i/\nu$ and the ratios b/a and Ω_o/Ω_i, where the latter is the quotient of the outer cylinder rotation rate to the inner as well as the boundary conditions on the ends. This

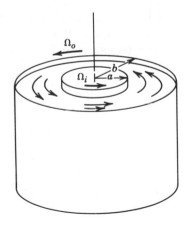

Figure 3-37 Sketch of flow between two rotating cylinders known as Taylor–Couette flow.

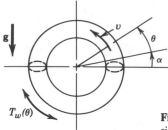

Figure 3-38 Thermal convection in a vertical one-dimensional fluid circuit. A model for a thermosiphon.

system exhibits a prechaos behavior of quasiperiodic oscillations before broad-band chaotic noise sets in. Other work includes that of Brandslater et al. (1983).

Closed-Loop Thermosiphon. It is curious, given the great amount of attention to the Lorenz attractor as a paradigm for convective flow chaos, that only a few attempts have been made to design an experiment that embodied all the assumptions in the Lorenz model. The experiment that comes very close to the Lorenz model is the flow of fluid in a circular channel under gravity. The relevance of this experiment to the Lorenz model has been pointed out by Hart (1984). Convectively driven flows are of interest as models for geophysical flows such as warm springs or groundwater flow through permeable layers in the earth's crust as well as applications for solar heating systems or reactor core cooling.

Early experiments by Bau and Torrance (1981) were performed in a rectangular loop thermosiphon. They derived equations that describe flow in a closed circular tube with gravity acting in the vertical plane, as shown in Figure 3-38. Essentially, all variables are assumed to be independent of the radial direction. The principal dependent variables are the circumferential velocity $v(t)$ and the temperature $T(\theta, t)$. A viscous wall stress is assumed to act in the fluid. Also, a prescribed wall temperature $T_w(\theta)$ is assumed with a linear cooling law proportional to $T - T_w$.

The basic equations are the balance of angular momentum for the fluid mass and a partial differential equation for the energy or heat balance law.

A buoyancy force or moment is introduced by assuming that the fluid density depends on the temperature,

$$\rho = \rho_0 [1 - \beta(T - T_0)] \qquad (3\text{-}3.16)$$

so that a net torque acts on the fluid proportional to

$$g\beta a \int_{-\pi}^{\pi} T(\theta)\cos(\theta + \alpha)\, d\theta \qquad (3\text{-}3.17)$$

where θ is defined in Figure 3-38.

In a method similar to that used in deriving the Lorenz equations (3-2.3), the temperature is expanded in a Fourier series. In this way, the partial differential equation for the heat balance is reduced to a set of ordinary differential equations.

Following Hart (1984), one writes

$$T(\theta) = \sum C_n(t)\cos n\theta + S_n(t)\sin n\theta \tag{3-3.18}$$

He shows that only the $n = 1$ thermal modes determine the dynamics. By redefining variables $x = v$, $y = C_1$, and $z = S_1 + R_a$, where R_a is similar to the Reynolds number, the resulting coupled first-order equations take the form

$$\dot{x} = P_r\left[-F(x) + (\cos\alpha)y - (\sin\alpha)(z - R_a)\right]$$

$$\dot{y} = -xz - y + R_a x \tag{3-3.19}$$

$$\dot{z} = xy - z$$

where $F(x)$ is a nonlinear friction law. To obtain the Lorenz equations, one sets $\alpha = 0$ and $F(x) = Cx$. The Lorenz limit corresponds to antisymmetric heating about the vertical. In their experiments, Bau and Torrance (1981) investigated the stability of flow but did not explore the chaotic regime. Given the close correspondence between Eqs. (3-3.19) and the Lorenz equations (3-2.3), it would appear natural that experimental exploration of the chaotic regime of the thermosiphon would be attempted. Another analysis of the relation between the Lorenz equations and fluid in a heated loop has been reported by Yorke et al. (1985).

Earlier experiments with a fluid convection loop by Creveling et al. (1975) did not report chaotic motions. However, recent experiments by Gorman et al. (1984) have reproduced some of the features of the Lorenz attractor. The working fluid was water and the apparatus consisted of a 75 cm diameter loop of 2.5 cm diameter Pyrex (glass) tubing. The bottom half was heated with electrical resistance tape while the top half was kept in a constant-temperature bath.

Pipe Flow Chaos. While closed-flow problems have captured the bulk of the attention vis-à-vis dynamical systems theory, open-flow problems are of great importance to engineering design. These include flows over air foils, boundary layers, jets and pipe flow. Recently, increased attention has been focused on applying the theory of chaotic dynamics to the laminar–turbulent transition problem in open-flow systems. One example is the experiment of Sreenivasan (1986) of Yale University who is studying intermittency in pipe flows. In this problem, low-velocity flow is laminar and steady, while for sufficiently high mean flow velocity the flow field becomes turbulent. At some critical velocity, the transition from laminar to turbulent appears to occur in intermittent bursts of turbulence. As the velocity increases, the fraction of time spent in the chaotic state increases until the flow is completely turbulent. Some observations of this phenomenon go back to Reynolds in 1883. The current focus of attention is to try to relate features of the intermittency, such as distribution of burst times, to dynamical theories of intermittency (e.g., see Pomeau and Manneville, 1980).

Fluid Drop Chaos. A simple system with which the reader can observe chaotic dynamics in one's home is the dripping faucet. This experiment is described by R. Shaw of the University of California–Santa Cruz in a monograph on chaos and information theory (1984). The experiment and sketch of experimental data are shown in Figure 3-39. The observable variable is the time between drops as measured with a light source and photocell, and the control variable is the flow rate from the nozzle. In Shaw's experiment, he measures a sequence of time intervals $\{T_n, T_{n+1}, T_{n+2}\}$ but does not measure the drop size or other physical properties of the drop such as shape. He and his students obtained periodic motion and period-doubling phenomena as well as chaotic behavior. Different maps of T_{n+1} versus T_n are obtained for different flow rates. The map in Figure 3-39 shows a classic one-dimensional parabolic map similar to the logistic map of Feigenbaum (1978). They also observed a more complicated map which is best represented in a three-dimensional phase space T_n versus T_{n+1} versus T_{n+2}. This is an example of using discrete data to construct a pseudo-phase-space and suggests that another dynamic variable should be observed (such as drop size) (see Chapter 4).

Surface Wave Chaos. It is well known that waves can propagate on the interface between two immiscible fluids under gravity (e.g., air on water). Such waves can be excited by vibrating a liquid in the vertical direction in the same way that one can parametrically excite vibrations in a pendulum. Subharmonic excitation of shallow water waves goes back to Faraday in 1831. An analysis of this phenomenon in the context of period doubling has been performed by a group at UCLA (Keolian et al., 1981). They looked at saltwater waves in an annulus of 4.8 cm mean radius with a cross section of 0.8×2.5 cm. The system is driven in the vertical direction by placing the annulus on an acoustic loudspeaker. By measuring the wave height versus time at several locations around the annulus, the UCLA group measured a subharmonic sequence before chaos that does not follow the classic period-

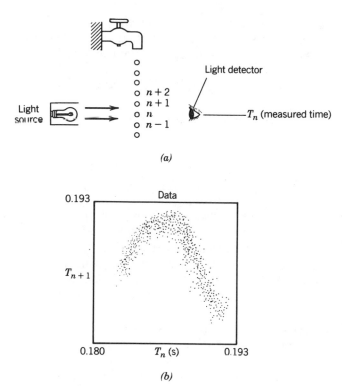

(a)

(b)

Figure 3-39 Experimental one-dimensional map for the time between drops in a dripping faucet [from Shaw (1984) with permission of Ariel Press, copyright 1984].

doubling sequence; for example, they observe resonant frequencies pf/m, where f is the driving frequency, for $m = 1, 2, 4, 12, 14, 16, 18, 20, 24, 28, 36$, which differs from the 2^n sequence of the logistic equation.

In another study of forced surface waves, Ciliberto and Gollub (1985) looked at a cylindrical dish of water with radius 6.35 cm with a depth of about 1 cm. They also use a loudspeaker excitation to explore regions of periodic and chaotic motion of the fluid height. In the region around 16 Hz, for example, they obtain chaotic wave motion for a vertical driving height of around 0.15 mm. They try to interpret the results in terms of nonlinear interaction between two linear spatial modes. A theoretical analysis of this problem has been done by Holmes (1986).

Chaos in Chemical Reactions. Rossler (1976) Hudson et al. (1984), have observed chaotic dynamics in a small reactor-diffusion system. Also, Schrieber et al. (1980) have observed similar behavior in two coupled stirred-cell reactors. If (x_1, y_1) represents the chemical concentration in one cell and (x_2, y_2) represents the concentration in another cell, a set of equations can be derived to model the dynamic behavior:

$$\dot{x}_1 = A - (B + 1)x_1 + x_1^2 y_1 + D_1(x_2 - x_1)$$

$$\dot{y}_1 = B_1 x_1 - x_1^2 y_1 + D_2(y_2 - y_1)$$

$$\dot{x}_2 = A - (B + 1)x_2 + x_2^2 y_2 - D_1(x_1 - x_2)$$ (3-3.20)

$$\dot{y}_2 = Bx_2 - x_2^2 y_2 + D_2(y_1 - y_2)$$

A now classic example of chemical chaos is the Belousov–Zhabotinski reaction in a stirred-flow reactor. Subharmonic oscillations and period doubling have been observed by Simoyi et al. (1982) (Figure 3.34). With the input chemical concentrations held fixed, the time history of the concentration of the bromide ion, one of the reaction chemicals, shows complex subharmonic behavior for different flow rates.

Light Wave Chaos.   Many papers have been published in the physics literature on chaotic behavior of laser systems as well as for the chaotic propagation of light through nonlinear optical devices. An extensive review of chaos in light systems has been written by Harrison and Biswas (1986). In elementary laser systems the nonlinearity originates from the fact that the system oscillates between at least two discrete energy levels. The simplest mathematical model for such systems involves three first-order equations for the electric field in the laser cavity, the population inversion, and the atomic polarization. These equations, known as the Maxwell–Block equations, are similar in structure to the Lorenz equations discussed in Chapters 1 and 3 (3-2.3). Chaotic phenomena in lasers have been observed in both the autonomous mode and the modulation mode.

The other class of problems discussed in Harrison and Biswas (1986) involves passive nonlinear optics. Here the index of refraction (speed of light in the medium) depends on the intensity of the light, for example, through the Kerr effect.

Bio Chaos.   One of the exciting aspects of the new mathematical models in nonlinear dynamics is the wide applicability of these paradigms to many different fields of science. Thus, it is no surprise that dynamic phenomena in biological systems that have exhibited periodic and chaotic motions, have been explained by some of the very same equations used in the electrical and mechanical sciences. We mention just two examples here.

Chaotic Heart Beats. Glass et al. (1983) have performed dynamic experiments on spontaneous beating in groups of cells from embryonic chick hearts. Without external stimuli, these oscillations have a period between 0.4 and 1.3 s. However, when periodic current pulses are sent into the group using microelectrodes, entrainment, quasiperiodicity, and chaotic motions have been observed.

A discussion of the relevance of nonlinear dynamics and chaotic models to ventricular fibrillation has recently been given by Goldberger et al. (1986). This paper contains a number of references on cardiac dynamics.

Nerve Cells. In a similar type of experiment, sinusoidal stimulation of a giant neuron in a marine mollusk by Hayashi et al. (1982) also showed evidence of chaotic behavior.

4

Experimental Methods in Chaotic Vibrations

Perfect logic and faultless deduction make a pleasant
theoretical structure, but it may be right or wrong:
The experimenter is the only one to decide, and he is
always right.
L. Brillouin, *Scientific Uncertainty and Information*, 1964

4.1 INTRODUCTION: EXPERIMENTAL GOALS

A review of physical systems that exhibit chaotic vibrations was presented in Chapter 3. In this chapter, we discuss some of the experimental techniques that have been used successfully to observe and characterize chaotic vibrations and strange attractors. To a great extent, these techniques are specific to the physical medium, for example, rigid body, elastic solid, fluid, or reacting medium. However, many of those measurements which are unique to chaotic phenomenon, such as Poincaré maps or Lyapunov exponents, are applicable to a wide spectrum of problems.

A diagram outlining the major components of an experiment is shown in Figure 4-1. In this example, the vibrating object is an elastic beam with either nonlinear boundary conditions or multiple equilibrium positions. Also, the source of the vibration is an electromagnetic shaker. In the case of an autonomous system, such as the Rayleigh–Benard convection cell, the source of instability is a prescribed temperature difference across the cell, and the nonlinearities reside in the convective terms in the acceleration of each fluid element.

The other major elements include *transducers* to convert physical variables into electronic voltages, a *data acquisition* and storage system, *graphical display* (such as an oscilloscope), and data analysis computer.

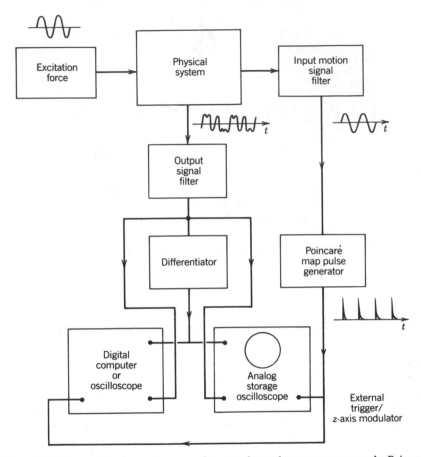

Figure 4-1 Diagram showing components of an experimental system to measure the Poincaré map of a chaotic physical system.

The techniques that must be mastered for experiments in chaotic vibrations depend to some extent on the goals that one sets up for the experimental study. These goals could include the following:

1. Establish existence of chaotic vibration in a particular physical system.
2. Determine critical parameters for bifurcations.
3. Determine criterion for chaos.
4. Map out chaotic regimes.
5. Measure qualitative features of chaotic attractor, for example, Poincaré maps.

6. Measure quantitative properties of attractor, for example, Fourier spectrum, Lyapunov exponent, probability density function, fractal dimension.

4.2 NONLINEAR ELEMENTS IN DYNAMICAL SYSTEMS

The phenomena of chaotic vibrations cannot occur if the system is linear. Thus, in performing experiments in chaotic dynamics, one should understand the nature of the nonlinearities in the system. To refresh one's memory, a linear system is one in which the principle of superposition is valid. Thus, if $x_1(t)$ and $x_2(t)$ are each possible motions of a given system, then the system is linear if the sum $c_1x_1(t) + c_2x_2(t)$ is also a possible motion. Another form of the superposition principle is more easily described in mathematical terms. Suppose the dynamics of a given system can be modeled by a set of differential or integral equations of the form

$$L[X] = f(t) \qquad (4\text{-}2.1)$$

and $X = (x_1, x_2, \ldots, x_k(t), \ldots, x_n)$ represents a set of independent dynamical variables that describe the system. Suppose the system is forced by two different input functions $f_1(t)$ and $f_2(t)$, with outputs $X_1(t)$ and $X_2(t)$. If the system is linear, the effect of two simultaneous inputs can easily be found:

$$L[c_1X_1 + c_2X_2] = c_1f_1(t) + c_2f_2(t) \qquad (4\text{-}2.2)$$

The only way that this property can hold is for the terms in the differential equation (4-2.1) to be to the first power X_1 or $\dot{X}_1$, and so on: hence the term *linear* system. Nonlinear systems involve the unknown functions in forms other than to the first power, that is, $x^2, x^3, \sin x, x^a, 1/(x^2 + b)$, or similar forms for the derivatives or integrals of the function, that is, $\dot{x}^2, [\int x\, dt]^2$.

Experimental nonlinearities can be created in many ways, some of them quite subtle. In mechanical or electromagnetic systems, nonlinearities can occur in the following forms:

(a) Nonlinear material or constitutive properties (stress versus strain, voltage versus current)
(b) Nonlinear acceleration or kinematic terms (e.g., centripetal or Coriolis acceleration terms)
(c) Nonlinear body forces
(d) Geometric nonlinearities

(a) Material Nonlinearities

Examples of material nonlinearities in mechanical and electrical systems include the following

Solid Materials. Nonlinear stress versus strain: (1) elastic (e.g., rubber) and (2) inelastic (e.g., steel beyond the yield point, plasticity, creep).

Magnetic Materials. Nonlinear magnetic field intensity **H** versus flux density **B**

$$\mathbf{B} = \mathbf{f}(\mathbf{H})$$

(e.g., ferromagnetic material iron, nickel, cobalt—hysteretic in nature).

Dielectric Materials. Nonlinear electric displacement **D** versus electric field intensity **E**

$$\mathbf{D} = \mathbf{f}(\mathbf{E})$$

(e.g., ferroelectric materials).

Electric Circuit Elements. Nonlinear voltage versus current

$$V = f(I)$$

(e.g., Zener and tunnel diodes, nonlinear resistors, field effect transistors (FET), metal oxide semiconductors (MOSFET)). Nonlinear voltage versus charge

$$V = g(Q)$$

(e.g., capacitors). Other material nonlinearities include nonlinear optical materials (e.g., lasers), heat flux–temperature gradient properties, nonlinear viscosity properties in fluids, voltage–current relations in electric arcs, and dry friction.

(b) Kinematic Nonlinearities

This type of nonlinearity occurs in fluid mechanics in the Navier–Stokes equations where the acceleration term includes a nonlinear velocity operator

$$v\frac{\partial v}{\partial x} \quad \text{or} \quad \mathbf{v} \cdot \nabla \mathbf{v}$$

which represents convective effects.

In particle dynamics, one often uses local coordinate systems to describe motion relative to some inertial reference frame. When the local frame rotates with angular velocity Ω relative to the large frame, the absolute acceleration is given by

$$\mathbf{A} = \mathbf{a} + \mathbf{A}_0 + \dot{\Omega} \times \mathbf{\rho} + \Omega \times \Omega \times \mathbf{\rho} + 2\Omega \times \mathbf{v} \qquad (4\text{-}2.3)$$

where $\mathbf{A}_0$ is the acceleration of the origin of the small frame relative to the reference, and $\mathbf{\rho}$ and $\mathbf{v}$ are the local position vector and velocity, respectively, of the particle. The last two terms are called the centripetal and Coriolis acceleration terms. The last three terms are *nonlinear* in the variables $\mathbf{\rho}$, $\mathbf{v}$, Ω.

For a rigid body in pure rotation, nonlinear terms appear in Euler's equations for the rotation dynamics:

$$M_x = I_x \frac{d\omega_x}{dt} - \left(I_z - I_y \right) \omega_y \omega_z$$

$$M_y = I_y \frac{d\omega_y}{dt} - \left(I_x - I_z \right) \omega_z \omega_x \qquad (4\text{-}2.4)$$

$$M_z = I_z \frac{d\omega_z}{dt} - \left(I_y - I_x \right) \omega_x \omega_y$$

where (M_x, M_y, M_z) are applied force moments and (I_x, I_y, I_z) are principal second moments of mass about the center of mass.

(c) Nonlinear Body Forces

Electromagnetic forces are represented as follows:

$$\text{Currents} \qquad F = \alpha I_1 I_2 \quad \text{or} \quad \beta I B$$

$$\text{Magnetization} \quad \mathbf{F} = \mathbf{M} \cdot \nabla \mathbf{B}$$

$$\text{Moving media} \quad \mathbf{F} = q\mathbf{v} \times \mathbf{B}$$

(Here I is current, $\mathbf{B}$ is the magnetic field, $\mathbf{M}$ is the magnetization, q represents charge, and $\mathbf{v}$ is the velocity of a moving charge.)

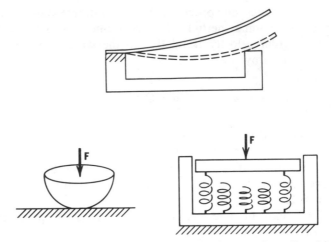

Figure 4-2 Examples of mechanical systems with geometric nonlinearities.

(d) Geometric Nonlinearities

Geometric nonlinearities in mechanics involve materials with linear stress–strain behavior but the geometry changes with deformation. One example is shown in Figure 4-2 where the constraint on the tip displacement of the cantilever depends on the displacement.

Another classic example is the contact of two smooth elastic bodies (called a Hertz contact). The force–displacement law for curved surfaces follows a nonlinear power law

$$F = c\delta^{3/2}$$

where δ is the relative approach of the two bodies.

Another classic example of a geometric nonlinearity is the elastica shown in Figure 4-3. In this problem, the material is linearly elastic but the large deformations produce a nonlinear force–displacement or moment–angle relation of the form

$$M = A\kappa$$

$$\kappa = \frac{u''}{\left[1 + (u')^2\right]^{3/2}}$$

where M is the bending moment, κ is the curvature of the neutral axis of the beam, and $u(x)$ is the transverse displacement of the beam. This

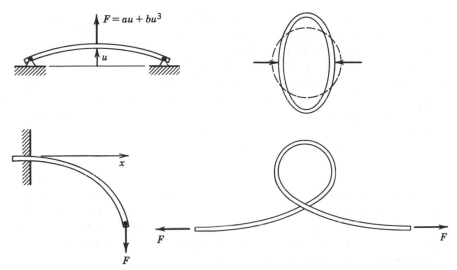

Figure 4-3   Examples of geometric nonlinearities in elastic structures.

problem is an interesting one for study of chaotic vibrations since the elastica can exhibit multiple equilibrium solutions (see Chapter 2). Cylindrical and spherical shells also exhibit geometric elastic nonlinearities (e.g., see Evensen, 1967).

4.3 EXPERIMENTAL CONTROLS

First and foremost, the experimenter in chaotic vibrations should have control over noise, both mechanical and electronic. If one is to establish chaotic behavior for a deterministic system, the noise inputs to the system must be minimized.

For example, mechanical experiments such as vibration of structures or autonomous fluid convection problems should be isolated from external laboratory or building vibrations. This can be accomplished by using a large-mass table with low-frequency air bearings. A low-cost solution is to work at night when building noise is at a minimum.

Second, one should build in the ability to control significant physical parameters in the experiments, such as forcing amplitude or temperature gradient. This is especially important if one wishes to observe bifurcation sequences such as period-doubling phenomena. Where possible, one should use continuous element controls and avoid devices with incremental or step changes in the parameters. In some problems, there is more than one dynamic motion for the same parameters. Thus, control over the initial state variables may also be important.

Control of the number of degrees of freedom is another consideration. For example, if one wishes to observe low-frequency oscillations of a structure, care should be taken to make sure other vibration modes are not excited. Other extraneous vibration modes can creep into the experiment from the boundary conditions that support or clamp the structure. This may call for securing the structure to a large-mass base.

Another factor is the number of significant figures required for accurate measurement. For example, to plot Poincaré maps from digitally sampled data, an 8 bit system may not be sensitive enough and one may have to go with 12 bit electronics or better. In some of our experiments on Poincaré maps, we have obtained better results from analog devices, such as a good analog storage oscilloscope, than an 8 bit digital oscilloscope especially as regards resolution of fine fractal structure in the maps.

Frequency Bandwidth

Most experiments in fluid, solid, or reacting systems may be viewed as infinite-dimensional continua. However, one often tries to develop a mathematical model with a few degrees of freedom to explain the major features of the chaotic or turbulent motions of the system. This is usually done by making measurements at a few spatial locations in the continuous system and by limiting the frequency bandwidth over which one observes the chaos. This is especially important if velocity measurements for phase plane plots are to be made from deformation histories. Electronic differentiation will amplify higher-frequency signals, which may not be of interest in the experiment. Thus, extremely good electronic filters are often required, especially ones that have little or no phase shift in the frequency band of interest.

4.4 PHASE SPACE MEASUREMENTS

It was pointed out in Chapter 2 that chaotic dynamics are most easily unraveled and understood when viewed from a phase space perspective. In particle dynamics, this means a space with coordinates composed of the position and velocity for each independent degree of freedom. In forced problems, time becomes another dimension. Thus, the periodic forcing of a two-degree-of-freedom oscillator with generalized positions $(q_1(t), q_2(t))$

has a phase space representation with coordinates $(q_1, \dot{q}_1, q_2, \dot{q}_2, \omega t)$, where ω is the forcing frequency.

If one measures displacement $q(t)$, a differentiation circuit is required. If velocity is measured, the phase space may be spanned by $(v, \int v \, dt)$, which calls for an integrator circuit. As noted above, in building integrator or differentiator circuits, care should be taken that the phase as well as the amplitude is not distorted within the frequency band of interest.

In electronic or electrical circuit problems, the current and voltage can be used as state variables. In fluid convection problems, temperature and velocity variables are important.

Pseudo-Phase-Space Measurements

In many experiments, one has access to only one measured variable $\{x(t_1), x(t_2), \ldots\}$ (where t_1 and t_2 are sampling times, not to be confused with Poincaré maps). When the time increment is uniform, that is, $t_2 = t_1 + \tau$ and so on, a pseudo-phase-space plot can be made using $x(t)$ and its past (or future) values:

$(x(t), x(t - \tau))$,   or $(x(t), x(t + \tau))$ two-dimensional phase space

$(x(t), x(t - \tau), x(t - 2\tau))$ three-dimensional phase space

One can show that a closed trajectory in a phase space in $(x, \dot{x})$ variables will be closed in the $(x(t), x(t - \tau))$ variables (one must connect the points when the system is digitally sampled) as shown in Figure 4-4. Likewise, chaotic trajectories in $(x, \dot{x})$ look chaotic in $(x(t), x(t - \tau))$ variables. The plots can be carried out after the experiment by a computer or one may perform on-line pseudo-phase-plane plots using a sample and hold circuit.

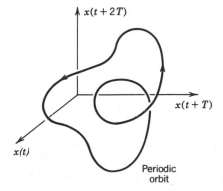

Periodic orbit

Figure 4-4 Periodic trajectory of a third-order dynamical system using pseudo-phase-space coordinates.

The one difficulty with pseudo-phase-space variables is taking a Poincaré map. For example, when there is a natural time scale, such as in forced periodic motion of a system with frequency ω, the sample time τ is usually chosen much smaller than the driving period; that is, $\tau \ll 2\pi/\omega \equiv T$. If τ is not an integer fraction of T, Poincaré maps may lose some of the fine fractal structure.

4.5 BIFURCATION DIAGRAMS

As discussed in Chapter 2, one of the signs of impending chaotic behavior in dynamical systems is a series of changes in the nature of the periodic motions as some parameter is varied. Typically, in a single-degree-of-freedom oscillator, as the control parameter approaches a critical value for chaotic motion, subharmonic oscillations appear. In the now classic "logistic equation," a series of period 2 oscillations appear [Eq. (1-3.6)]. The phenomenon of sudden change in the motion as a parameter is varied is called a *bifurcation*. A sample experimental bifurcation diagram is shown in Figure 4-5. Such diagrams can be obtained experimentally by time sampling the motion as in a Poincaré map and displaying the output on an oscilloscope as shown in Figure 4-5. Here the value of the control parameter, for example, a forcing amplitude or frequency, is plotted on the horizontal axis and the time-sampled values of the motion are plotted on the vertical axis. This diagram actually represents a series of experiments, where each value of the control parameter is an experiment. When the control parameter can be varied automatically, such as by a computer and digital-to-analog device, the diagram can be obtained quite rapidly. Care must be taken, however, to make sure transients have died out after each change in the control parameter.

In the bifurcation diagram of Figure 4-5, the continuous horizontal lines represent periodic motions of various subharmonics. The values in the dashed line areas represent chaotic regions. The boundary between chaotic and periodic motions can clearly be seen in this diagram.

When this is automated, one must be careful not to mistake a quasiperiodic motion for a chaotic motion. A phase plane Poincaré map is still very useful for distinguishing between quasiperiodic and chaotic motions.

4.6 EXPERIMENTAL POINCARÉ MAPS

Poincaré maps are one of the principal ways of recognizing chaotic vibrations in low-degree-of-freedom problems (see Table 2-2). We recall that the dynamics of a one-degree-of-freedom forced mechanical oscillator or

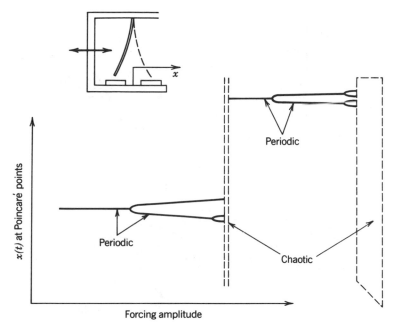

Figure 4-5 Experimental bifurcation diagram for the vibration of a buckled beam: Poincaré map samples of bending displacement versus amplitude of forcing vibration.

L–R–C circuit may be described in a three-dimensional phase space. Thus, if $x(t)$ is the displacement, $(x, \dot{x}, \omega t)$ represents a point in a cylindrical phase space where $\phi = \omega t$ represents the phase of the periodic forcing function. A Poincaré map for this problem consists of digitally sampled points in this three-dimensional space, for example, $(x(t_n), \dot{x}(t_n), \omega t_n = 2\pi n)$. As discussed in Chapter 2, this map can be thought of as slicing a torus (see Figure 4-6).

Experimentally this can be done in several ways. If one has a storage oscilloscope, the Poincaré map is obtained by intensifying the image on the screen at a certain phase of the forcing voltage (sometimes called *z-axis modulation*) (Figure 4-1). In our laboratory, we were able to generate a 5–10 V pulse of 1–2 μs duration when the forcing function reached a certain phase:

$$\omega t_n = \phi_0 + 2\pi n \qquad (4\text{-}6.1)$$

This pulse was then used to intensify a phase plane image, $(x(t), \dot{x}(t))$, using two vertical amplifiers as in Figure 4-7.

One can also use a digital oscilloscope in an external sampling rate mode with the same narrow pulse signal used for the analog oscilloscope. A

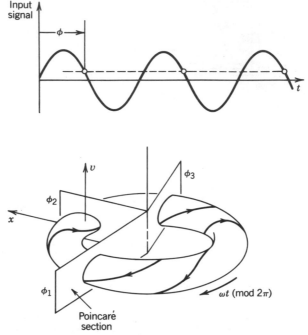

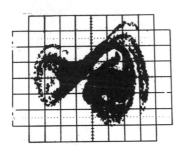

Figure 4-6 *Top*: Poincaré map sampling times at constant phase of forcing function. *Bottom*: Geometric interpretation of Poincaré sections in the three-dimensional phase space.

Figure 4-7 Example of an experimental Poincaré map for periodic forcing of a buckled beam.

similar technique can be employed using an analog-to-digital (A-D) signal converter by storing the sampled data in a computer for display at a later time. The important point here is that the sampling trigger signal must be exactly synchronous with the forcing function.

Poincaré Maps—Change of Phase. As noted in Chapter 2, chaotic phase plane trajectories can often be unraveled using the Poincaré map by taking

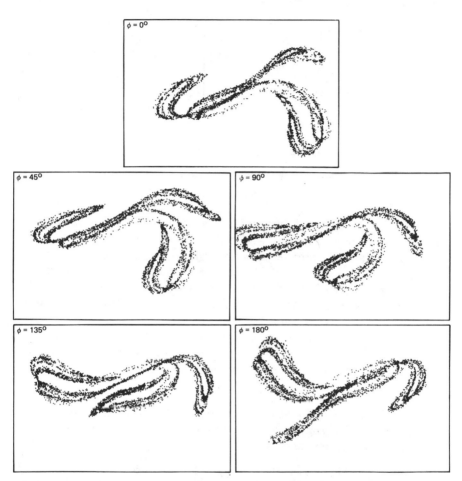

Figure 4-8 Poincaré maps of a chaotic attractor for a buckled beam for different phases of the forcing function.

a set of pictures for different phases ϕ_0 in Eq. (4-6.1) (see Figure 4-8). This is tantamount to sweeping the Poincaré plane in Figure 4-6. While one Poincaré map can be used to expose the fractal nature of the attractor, a complete set of maps varying ϕ_0 from 0 to 2π is sometimes needed to obtain a complete picture of the attractor on which the motion is riding.

A series of pictures of various cross sections of a chaotic torus motion in a three-dimensional phase space is shown in Figure 4-8. Note the symmetry in the $\varphi = 0°$ and $180°$ maps for the special case of the buckled beam (Figure 4-5).

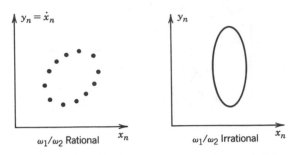

Figure 4-9   Poincaré map of a motion with two harmonic signals with different frequencies.

Poincaré Maps—Effect of Damping.   If a system does not have sufficient damping, the chaotic attractor will tend to fill up a section of phase space uniformly and the Cantor set structure, which is characteristic of strange attractors, will not be evident. An example of this is shown in Figure 2-11 for the vibration of a buckled beam. A comparison of low- and high-damping Poincaré maps shows that adding damping to the system can sometimes bring out the fractal structure.

On the other hand, if the damping is too high, the Cantor set sheets can appear to collapse onto one curve. The effect of damping on Poincaré maps and fractal dimension is discussed in Chapter 6.

Poincaré Maps—Quasiperiodic Oscillations.   Often what appears to be chaotic may very simply be a superposition of two incommensurate harmonic motions, for example,

$$x(t) = A\cos(\omega_1 t + \phi) + B\cos(\omega_2 t + \phi_2) \qquad (4\text{-}6.2)$$

where ω_1/ω_2 is irrational. One can use one frequency to sample a Poincaré map, for example, $\omega_1 t_n = 2\pi n$. Then the phase plane points $(x(t_n), \dot{x}(t_n))$ will fill in an elliptically shaped closed curve (Figure 4-9). If ω_1/ω_2 is rational, a finite set of points will be seen in the Poincaré map. The case of ω_1/ω_2 irrational can be thought of as motion on a torus or doughnut-shaped figure in a three-dimensional phase space.

When three or more incommensurate frequencies are present one may not see a nice closed curve in the Poincaré map and the Fourier spectrum should be used. The difference between chaotic and quasiperiodic motion can also be detected by taking the Fourier spectrum of the signal. A quasiperiodic motion will have a few well-pronounced peaks as shown in Figure 4-10. Chaotic signals often have a broad spectrum of Fourier components as in Figure 2-7.

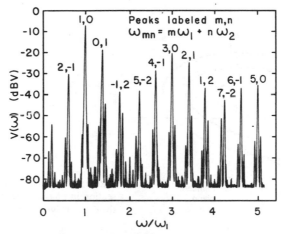

Figure 4-10 Fourier spectrum of an experimental electronic signal from a circuit with a nonlinear inductor [from Bryant and Jeffries (1984) with permission of The American Physical Society, copyright 1984]. Frequency components are linear combinations of two frequencies.

Position Triggered Poincaré Maps

When one does not have a natural time clock, such as a periodic forcing function, then more sophisticated techniques must be used to get a Poincaré map. (See also Henon, 1982.)

Suppose we imagine the motion as a trajectory in a three-dimensional space with coordinates (x, y, z). To construct a Poincaré map, we intercept a trajectory with a plane defined by

$$ax + by + cz = d \qquad (4\text{-}6.3)$$

as shown in Figure 4-11. The Poincaré map consists of those points in the plane for which the trajectory penetrates the plane with the same sense [i.e., if we define a front and back to the plane (4-6.3), we collect points only on trajectories that penetrate the plane from front to back or back to front, but not both ways].

Experimentally, one can do this by using a mechanical or electronic *level detector*. Examples of position triggered Poincaré maps are discussed below.

In the impact oscillator shown in Figure 4-12, there are three convenient state variables: the position x, velocity v, and phase of the driving signal $\phi = \omega t$. If one triggers on the position when the mass hits the elastic constraint, the Poincaré map becomes a set of values $(v_n^{\pm}, \omega t_n)$, where $v_n^{\pm}$ is the velocity before or after impact and t_n is the time of impact. Here the points in the map can be plotted in a cylindrical space where $0 < \omega t_n < 2\pi$.

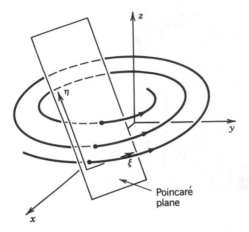

Figure 4-11 General Poincaré surface of section in phase space for the motion of a third-order dynamical system.

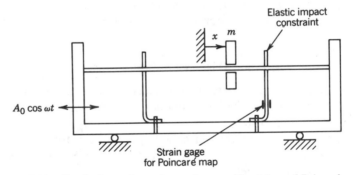

Figure 4-12 Sketch of experimental setup for a position triggered Poincaré section.

An example of the experimental technique to obtain a (v_n, ϕ_n) Poincaré map is shown in Figure 4-12. When the mass hits the position constraint, a sharp signal is obtained from a strain gauge or accelerometer. This sharp signal can be used to trigger a data storage device (such as a storage or digital oscilloscope) to store the value of the velocity of the particle. (In the case shown in Figure 4-12, a linear variable differential transformer —LVDT—is used to measure position, and this signal is electronically differentiated to get the velocity.) To obtain the phase ϕ_n modulo 2π, we generate a periodic ramp signal in phase with the driving signal where the minimum value of zero corresponds to $\phi = 0$, and the maximum voltage of the ramp corresponds to the phase $\phi = 2\pi$. The impact-generated sharp spike voltage is used to trigger the data storage device and store the value of the ramp voltage along with the velocity signal before or after impact. A

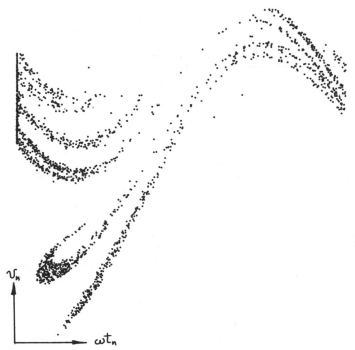

Figure 4-13   Position triggered Poincaré map for an oscillating mass with impact constraints (Figure 4-12).

Poincaré map for a mass bouncing between two elastic walls using this (v_n, ϕ_n) technique is shown in Figure 4-13.

Another example of this kind of Poincaré map is shown in Figure 4-14 for the chaotic vibrations of a motor. In this problem, the motor has a nonlinear torque–angle relation created by a dc current in one of the stator poles, and the permanent magnet rotor is driven by a sinusoidal torque created by an ac current in an adjacent coil. The equation of motion for this problem is

$$J\ddot{\theta} + \gamma\dot{\theta} + \kappa \sin\theta = F_0 \cos\theta \cos\omega t \qquad (4\text{-}6.4)$$

To obtain a Poincaré map, we choose a plane in the three-dimensional space $(\theta, \dot{\theta}, \omega t)$, where $\theta = 0$ (Figure 4-14). This is done experimentally by using a slit in a thin disk attached to the rotor and using a light-emitting diode and detector to generate a voltage pulse every time the rotor passes through $\theta = 0$ (see Figure 4-14). This pulse is then used to sample the velocity and measure the time. The data can be displayed directly on a

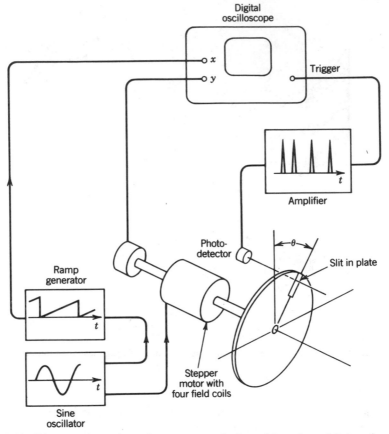

Figure 4-14 Diagram of experimental apparatus to obtain position triggered Poincaré maps for a periodically forced rotor with a nonlinear torque–angle relation.

storage oscilloscope or, using a computer, can be replotted in polar coordinates as shown in Figure 4-15.

Another variation of the method of Poincaré sections is to sample data when some variable attains a *peak* value. This has been used by Bryant and Jeffries (1984b) of the University of California–Berkeley. They examine the dynamics of a circuit with a nonlinear hysteretic iron core inductor shown in Figure 4-16. (The nonlinear properties are related to the ferromagnetic material in the inductor.) They sample the current in the inductor $I_L(t)$ as well as the driving voltage $V_s(t)$, when $V_L = 0$. This is tantamount to measuring the *peak* value of the flux in the inductor φ. This is because $V_L = -\dot{\varphi}$, where φ is the magnetic flux in the inductor, and $\varphi = \hat{\varphi}(I)$, so

Figure 4-15 Position triggered Poincaré map for chaos in a nonlinear rotor (see Figure 4-14).

that when $\dot{\varphi} = 0$, the flux is at a maximum or minimum. The Poincaré map is then a collection of pairs of points (V_{sn}, I_{Ln}) which can be displayed on a storage or digital oscilloscope.

Construction of One-Dimensional Maps from Multidimensional Attractors

There are a number of physical and numerical examples where the attracting set appears to have a sheetlike behavior in some three-dimensional phase space as illustrated in Figure 4-17. [The Lorenz equations (1-3.9) are such an example.] This often means that a Poincaré section, obtained by measuring the sequence of points that pierce a plane transverse to the attractor, will appear as a set of points along some one-dimensional line. This suggests that if one could parameterize these points along the line by a

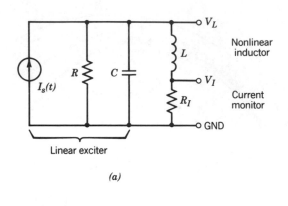

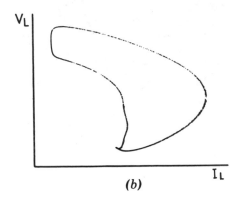

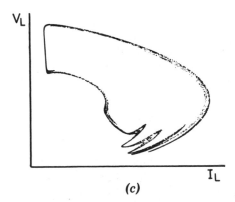

Figure 4-16 Peak amplitude-generated Poincaré maps for a circuit with nonlinear inductance [from Bryant and Jeffries (1984) with permission of The American Physical Society, copyright 1984].

140

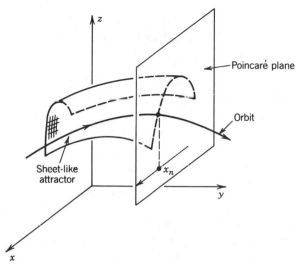

Figure 4-17 Construction of a one-dimensional return map in a three-dimensional phase space.

variable x, it would be possible that a function exists that relates x_{n+1} and x_n:

$$x_{n+1} = f(x_n)$$

The function (called a return map) may be found by simply plotting x_{n+1} versus x_n. One example is the experiments of Shaw (1984) on the dripping faucet shown in Figure 3-39 or the nonlinear circuit in Figure 3-32. (See also Simoyi et al., 1982). The existence of such a function $f(x)$ implies that the mathematical results for one-dimensional maps, such as period doubling and Feigenbaum scaling, may be applicable to the more complex physical problem in explaining, predicting, or organizing experimental observations.

For some problems, the function $f(x)$, when it exists, appears to cross itself or is tangled. This may suggest that the mapping function can be untangled by plotting the dynamics in a higher dimensional embedding space using three successive values of the Poincaré sampled data $[x(t_n), x(t_{n+1}),$ and $x(t_{n+2})]$. The three-dimensional nature of the relationship can sometimes be perceived by changing the projection of the three-dimensional curve onto the plane of a graphics computer monitor. This may suggest a special two-dimensional map of the form

$$x_{n+2} = f(x_{n+1}, x_n)$$

or

$$x_{n+1} = f(x_n, y_n)$$
$$y_{n+1} = Ax_n$$

This form is similar to the Henon map (1-3.8). This method has been used successfully by Van Buskirk and Jeffries (1985) in their study of circuits with $p-n$ junctions and by Brorson et al. (1983) who studied a sinusoidally driven resistor–inductor circuit with a varactor diode.

Double Poincaré Maps

So far we have only talked of Poincaré maps for third-order systems, such as a single degree of freedom with external forcing. But what about higher-order systems with motion in a four- or five-dimensional phase space? For example, a two-degree-of-freedom autonomous aeroelastic problem would have motion in a four-dimensional phase space (x_1, v_1, x_2, v_2), or if $x_1 \equiv x$, $(x(t), x(t - \tau), x(t - 2\tau), x(t - 3\tau))$. A Poincaré map triggered on one of the state variables would result in a set of points in a three-dimensional space. The fractal nature of this map, if it exists, might not be evident in three dimensions and certainly not if one projects this three-dimensional map onto a plane in two of the remaining variables.

A technique to observe the fractal nature of a three-dimensional Poincaré map of a fourth-order system has been developed in our laboratory which we call a *double Poincaré section* (see Figure 4-18). This technique enables one to slice a finite-width section of the three-dimensional map in order to uncover fractal properties of the attractor and hence determine if it is "strange." (See Moon and Holmes, 1985.)

We illustrate this technique with an example derived from the forced motion of a buckled beam. In this case, we examine a system with two incommensurate driving frequencies. The mathematical model[1] has the form

$$\dot{x} = y$$
$$\dot{y} = -\gamma y + F(x) + f_1 \cos \theta_1 + f_2 \cos(\theta_2 - \phi_0)$$
$$\dot{\theta}_1 = \omega_1$$
$$\dot{\theta}_2 = \omega_2$$

$$(4-6.5)$$

[1] In multidimensional dynamical systems, such as fluid-thermal problems, one important route to chaos is the occurrence of two limit cycle oscillations (Hopf bifurcations) resulting in quasiperiodic motion. This route to chaos was discussed in Chapter 2. The dynamics of this motion have been modeled by flow on a torus and the resulting Poincaré sections become closed circular arcs. Despite the importance of quasiperiodic oscillation to chaotic dynamics, very little has been done in other systems aside from fluids. Thus, we decided to explore quasiperiodic oscillations in a nonlinear structure such as a buckled beam.

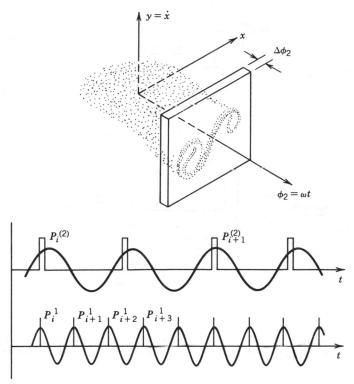

Figure 4-18 *Top*: Single Poincaré map dynamical system; finite width slice of second Poincaré section. *Bottom*: Poincaré sampling voltages for a second-order oscillator with two harmonic driving functions.

The experimental apparatus for a double Poincaré section is shown in Figure 4-19. The driving signals were produced by identical signal generators and were added electronically. The resulting quasiperiodic signal was then sent to a power amplifier which drove the electromagnetic shaker.

The first Poincaré map was generated by a 1 µs trigger pulse synchronous with one of the harmonic signals. The Poincaré map (x_n, v_n) using one trigger results in a fuzzy picture with no structure, as shown in Figure 4-20a. To obtain the second Poincaré section, we trigger on the second phase of the driving signal. However, if the pulse width is too narrow, the probability of finding points coincident with the first trigger is very small. Thus, we set the second pulse width 1000 times the first, at 1 ms. The second pulse width represents less than 1% of the second frequency phase of 2π. The (x, v) points were only stored when the first pulse was *coincident* with the second, as shown in Figure 4-18. This was accomplished using a

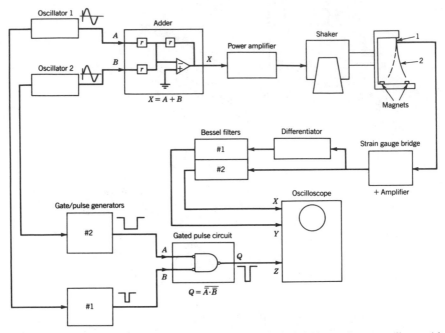

Figure 4-19 Sketch of experimental apparatus to obtain Poincaré map for an oscillator with two driving frequencies [from Moon and Holmes (1985) with permission of Elsevier Science Publishers, copyright 1985]. Note: Strain gauges—1; steel beam—2.

digital circuit with a logical NAND gate. Because of the infrequency of the simultaneity of both events, a map of 4000 points took more than 10 h compared to 8–10 min to obtain a conventional Poincaré map for driving frequencies less than 10 Hz.

The experimental results using this technique are shown in Figure 4-20 which compares a single with a double Poincaré map for the two-frequency forced problem. The single map is fuzzy while the double section reveals a fractal-like structure characteristic of a strange attractor.

One can of course generalize this technique to five- or higher-dimensional phase space problems. However, the probability of three or more simultaneous events will be very small unless the frequency is order of magnitudes higher than 1–10 Hz. Such higher-dimensional maps may be useful in nonlinear circuit problems.

This technique can be used in numerical simulation and has been employed by Lorenz (1984) to examine a strange attractor in a fourth-order system of ordinary differential equations. Kostelich and Yorke (1985) have also employed this method to study the dynamics of a kicked or pulsed

Figure 4-20 (*a*) Single Poincaré map of a nonlinear oscillator with two driving frequencies.

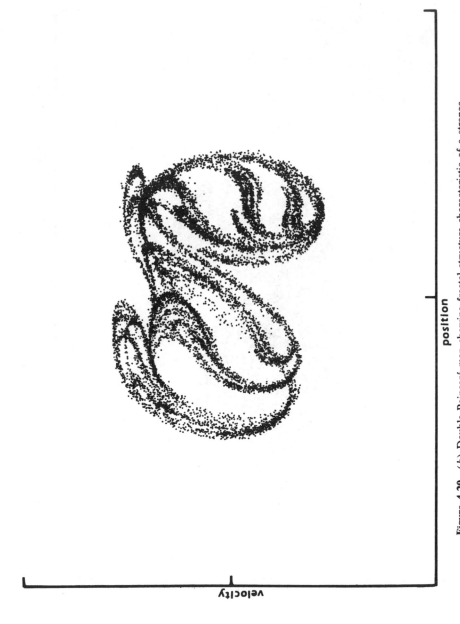

position

velocity

Figure 4-20 (*b*) Double Poincaré map showing fractal structure characteristic of a strange attractor.

146

double rotor. They call the method "Lorenz cross sections" (see also Kostelich et al., 1987).

4.7 QUANTITATIVE MEASURES OF CHAOTIC VIBRATIONS

Poincaré maps and phase plane portraits, when they can be obtained, can often provide graphic evidence for chaotic behavior and the fractal properties of strange attractors. However, quantitative measures of chaotic dynamics are also important and in many cases are the only hard evidence for chaos. The latter is especially true for systems with extreme frequencies 10^6–10^9 (as in laser systems) in which Poincaré maps may be difficult or impossible to capture. In addition, there are systems with many degrees of freedom where the Poincaré map will not reveal the fractal structure of the attractor section on double or multiple Poincaré maps) or the damping is so low that the Poincaré map shows no structure but looks like a cloud of points.

At this time in the development of the field there are three principal measures of chaos and another of emerging importance:

(a) Fourier distribution of frequency spectra
(b) Fractal dimension of chaotic attractor
(c) Lyapunov exponents
(d) Invariant probability distribution of attractor

It should be pointed out that while phase plane pictures and Poincaré maps can be obtained directly from electronic laboratory equipment, the above measures of chaos require a computer to analyze the data, with the possible exception of the frequency spectrum measurement. Electronic spectrum analyzers can be obtained but they are often expensive, and one might be better off investing in a laboratory micro or minicomputer that has the capability to perform other data analyses besides Fourier transforms.

If one is to analyze the data from chaotic motions digitally, then usually an *analog-to-digital converter* will be required as well as some means of storing the data. For example, the digitized data can be stored in a buffer in the electronic A-D device and then transmitted directly or over phone lines to a computer. Another option is a *digital oscilloscope* which performs the A-D conversion, displays the data graphically on the oscilloscope, and stores the data on a floppy disk. The latter method is often limited to eight 4000 bit records.

Finally, if one has the funds, one can store the output from the A-D converter directly onto a hard disk for direct transfer to a laboratory computer.

(a) Frequency Spectra—FFT

This is by far the most popular measure because the idea of decomposing a nonperiodic signal into a set of sinusoidal or harmonic signals is widely known among scientists and engineers. The assumption made in this method is that the periodic or nonperiodic signal can be represented as a synthesis of sine or cosine signals

$$f(t) = \frac{1}{2\pi} \int_{\Gamma} F(\omega) e^{i\omega t} \, d\omega \qquad (4\text{-}7.1)$$

where $e^{i\omega t} = \cos \omega t + i \sin \omega t$.

Since $F(\omega)$ is often complex, the absolute value $|F(\omega)|$ is used in graphical displays. In practice, one uses an electronic device or computer to calculate $|F(\omega)|$ from input data from the experiment while varying some parameter in the experiment (see Chapter 2, Routes to Chaos). When the motion is periodic or quasiperiodic, $|F(\omega)|$ shows a set of narrow spikes or lines indicating that the signal can be represented by a discrete set of harmonic functions $\{e^{\pm i\omega_k t}\}$, where $k = 1, 2, \ldots$. Near the onset of chaos, however, a continuous distribution of frequency appears, as shown in Figure 4-21a, and in the fully chaotic regime, the continuous spectrum may dominate the discrete spikes.

In general, the function $F(\omega)$ is a complex function of ω and to represent certain classes of signals $f(t)$, the integration (4-7.1) must be carried out along a path Γ in the complex ω plane. Numerical calculation of $F(\omega)$, given $f(t)$, can often be very time consuming even on a fast computer. However, most modern spectrum analyzers use a discrete version of (4-7.1) along with an efficient algorithm called the fast Fourier transform (FFT). Given a set of data sampled at discrete even time intervals $\{ f(t_k) = f_0, f_1, f_2, \ldots, f_k, \ldots, f_N\}$, the discrete time FFT is defined by the formula

$$T(J) = \sum_{I=1}^{N} f(I) e^{-2\pi i (I-1)(J-1)/N} \qquad (4\text{-}7.2)$$

where I and J are integers.

Several points should be made here which may appear obvious. First, the signal $f(t)$ is time sampled at a fixed time interval τ_0; thus, information is

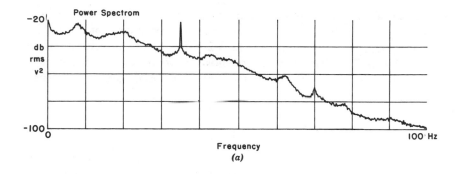

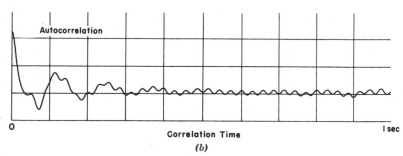

Figure 4-21 (*a*) Fourier spectrum of a chaotic signal. (*b*) Autocorrelation function of a chaotic signal.

lost for frequencies above $1/2\tau_0$. Second, only a finite set of points are used in the calculation, usually $N = 2^n$, and some built-in FFT electronics only do $N = 512$ or 1028 points. Thus, information is lost about very low frequencies below $1/N\tau_0$. Finally, the representation (4-7.2) having no information about $F(t)$ before $t = t_0$ or after $t = t_N$ essentially treats $f(t)$ as a periodic function. In general, this is not the case and since $f(t_0) \neq f(t_N)$, the Fourier representation treats this as a discontinuity which adds spurious information into $F(\omega)$. This is called *aliasing error* and methods exist to minimize its effect in $F(\omega)$. The reader using the FFT should be aware of this, however, when interpreting Fourier spectra about nonperiodic signals and should consult a signal processing reference for more information about FFTs.

Autocorrelation Function.   Another signal processing tool that is related to the Fourier transform is the autocorrelation function given by

$$A(\tau) = \int_0^\infty x(t)x(t + \tau)\, dt$$

When a signal is chaotic, information about its past origins is lost. This means that $A(\tau) \to 0$ as $\tau \to \infty$, or the signal is only correlated with its recent past. This is illustrated in Figure 4-21b for the chaotic vibrations of a buckled beam. The Fourier spectrum shows a broad band of frequencies, while the autocorrelation function has a peak at the origin $\tau = 0$, and drops off rapidly with time.

(b) Fractal Dimension

I will not go into too many technical details about fractal dimensions since Chapter 6 is devoted entirely to this topic. However, the basic idea is to characterize the "strangeness" of the chaotic attractor. If one looks at a Poincaré map of a typical low-dimensional strange attractor, as in Figure 4-8, one sees sets of points arranged along parallel lines. This structure persists when one enlarges a small region of the attractor. As noted in Chapter 2, this structure of the strange attractor differs from periodic motions (just a finite set of Poincaré points) or quasiperiodic motion which in the Poincaré map becomes a closed curve. In the Poincaré map, one can say that the dimension of the periodic map is zero and the dimension of the quasiperiodic map is one. The idea of the fractal dimension calculation is to attach a measure of dimension to the Cantor-like set of points in the strange attractor. If the points uniformly covered some area on the plane, we might say its dimension was close to two. Because the chaotic map in Figure 4-8 has an infinite set of gaps, its dimension is between one and two —thus the word fractal dimension.

In general, the set of Poincaré points in a strange attractor does not cover an integer-dimensional subspace (in Figure 4-8 this subspace is a plane).

Another use for the fractal dimension calculation is to determine the lowest order phase space for which the motion can be described. For example, in the case of some preturbulent convective flows in a Rayleigh–Benard cell (see Figure 3-1), the fractal dimension of the chaotic attractor can be calculated from some measure of the motion $\{x(t_n) \equiv x_n\}$ (see Malraison et al., 1983). From $\{x_n\}$, pseudo-phase-spaces of different dimension can be constructed (see Section 4.4). Using a computer algorithm, the fractal dimension d was found to reach an asymptotic $d = 3.5$ when the dimension of the pseudo-phase-space was four or larger. This suggests that a low-order approximation of the Navier–Stokes equation may be used to model this motion.

The reader is referred to Chapter 6 for further details. Although there are questions about the ability to calculate fractal dimensions for attractors of dimensions greater than four or five, this technique has gained increasing acceptance among experimentalists especially for low-dimensional chaotic attractors. If this trend continues, in the future, it is likely that electronic computing instruments will be available commercially that automatically calculate fractal dimension in the same way as FFTs are done at present.

(c) Lyapunov Exponents

Chaos in dynamics implies a sensitivity of the outcome of a dynamical process to changes in initial conditions. If one imagines a set of initial conditions within a sphere or radius ϵ in phase space, then for chaotic motions trajectories originating in the sphere will map the sphere into an ellipsoid whose major axis grows as $d = \epsilon e^{\lambda t}$, where $\lambda > 0$ is known as a *Lyapunov exponent*. (Lyapunov was a great Russian mathematician and mechanician 1857–1918.)

A number of experimenters in chaotic dynamics have developed algorithms to calculate the Lyapunov exponent μ. For regular motions $\lambda \leq 0$, but for chaotic motion $\lambda > 0$. Thus, *the sign of λ is a criterion for chaos.* The measurement involves the use of a computer to process the data. Algorithms have been developed to calculate λ from the measurement of a single dynamical variable $x(t)$ by constructing a pseudo-phase-space (e.g., see Wolf, 1984).

A more precise definition of Lyapunov exponents and techniques for measuring them is given in Chapter 5.

(d) Probability or Invariant Distributions

If a nonlinear dynamical system is in a chaotic state, precise prediction of the time history of the motion is impossible because small uncertainties in the initial conditions lead to divergent orbits in the phase space. If damping is present, we know that the chaotic orbit lies somewhere on the strange attractor. Failing specific knowledge about the whereabouts of the orbit, there is increasing interest in knowing the probability of finding the orbit somewhere on the attractor. One suggestion is to find a probability density in phase space to provide a statistical measure of the chaotic dynamics. There is some mathematical and experimental evidence that such a distribution does exist and that it does not vary with time.

To measure this distribution function, one time samples the motion at a number of points large enough to believe that the chaotic trajectory has visited most regions of the attractor. This minimum number can be determined by observing a Poincaré map to see when the attractor takes shape and when the Poincaré points fill in the different sections of the map. The

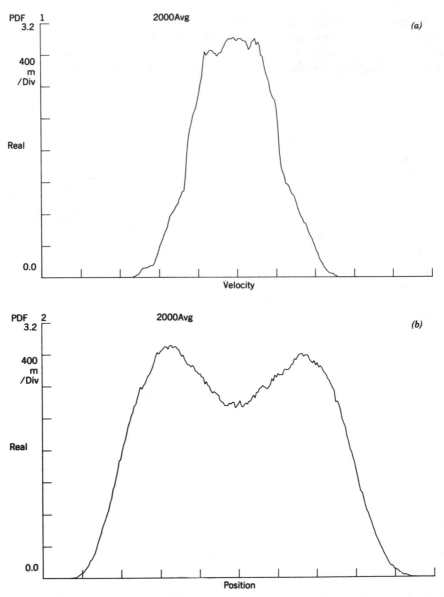

Figure 4-22 Experimental probability density function for chaotic vibration of a buckled beam averaged in time over many thousands of forcing periods. *Top*: Distribution of velocities of the beam tip; *Bottom*: Distribution of positions of the beam tip.

152

phase space is then partitioned into cells and the number of time-sampled points in each cell is counted using a computer.

An example of a probability density distribution for chaotic vibrations of the buckled beam problem is shown in Figure 4-22. Here we have projected the distribution onto the position axis and the velocity axis.

The distribution of velocities shows a shape similar to a classic Gaussian bell shaped curve (Figure 4-22a). The distribution of displacements on the other hand shows two peaks near the two potential wells (Figure 4-22b). This distribution is similar to that of a randomly excited two-well potential oscillator (Soong, 1973). It suggests that it might be possible to calculate probability density functions for deterministic chaotic systems using techniques from random vibration theory.

The usefulness of probability distribution for chaotic vibrations is similar to that for random vibrations (e.g., see Soong, 1973, or Lin, 1976). If the probability distribution can be determined for a chaotic system, one can calculate the mean square amplitude, mean zero crossing times, and probabilities of displacements, voltages, or stresses exceeding some critical value. However, much remains to be done on this subject both at the mathematical and experimental levels.

The use of probabilistic methods of analysis in chaotic vibrations has been developed by C. S. Hsu and coworkers at the University of California at Berkeley (Hsu, 1981, 1987; Hsu and Kim, 1985; Kreuzer, 1985 [now at Stuttgart]). This method, which divides the phase space into many cells, uses ideas from the theory of Markov processes. The method seems suited for the coming age of supercomputers and may become more widely known if it can be implemented in a parallel processing mode.

5

Criteria for Chaotic Vibrations

*But you will ask, how could a uniform chaos coagulate at
first irregularly in heterogeneous veins or masses to cause
hills — Tell me the cause of this, and the answer will
perhaps serve for the chaos.*
Isaac Newton, *On Creation* — from a letter circa 1681

5.1 INTRODUCTION

In this chapter, we study how the parameters of a dynamical system
determine whether the motion will be chaotic or regular. This is analogous
to finding the critical velocity in viscous flow of fluids above which steady
flow becomes turbulent. This velocity, when normalized by multiplying by a
characteristic length and dividing by the kinematic viscosity of the fluid, is
known as the critical *Reynolds number*, Re. A reliable theoretical value for
the critical Re has eluded engineers and physicists for over a century and
for most fluid problems experimental determination of $(Re)_{crit}$ is necessary.
In like manner, the determination of criteria for chaos in mechanical or
electrical systems in most cases must be found by experiment or computer
simulation. For such systems the search for critical parameters for deterministic chaos is a ripe subject for experimentalists and theoreticians
alike.

Despite the paucity of experimentally verified theories for the onset of
chaotic vibrations, there are some notable theoretical successes and some
general theoretical guidelines.

We distinguish between two kinds of criteria for chaos in physical
systems: a predictive rule and a diagnostic tool. A *predictive* rule for

chaotic vibrations is one that determines the set of input or control parameters that will lead to chaos. The ability to predict chaos in a physical system implies either that one has some approximate mathematical model of the system from which a criterion may be derived or that one has some empirical data based on many tests.

A *diagnostic* criterion for chaotic vibrations is a test that reveals if a particular system was or is in fact in a state of chaotic dynamics based on measurements or signal processing of data from the time history of the system.

We begin with a review of empirically determined criteria for specific physical systems and mathematical models which exhibit chaotic oscillations (Section 5.2). These criteria were determined by both physical and numerical experiments. We examine such cases for two reasons. First, it is of value for the novice in this field to explore a few particular chaotic systems in detail and to become familiar with the conditions under which chaos occurs. Such cases may give clues to when chaos occurs in more complex systems. Second, in the development of theoretical criteria, it is important to have some test case with which to compare theory with experiment.

In Section 5.3 we present a review of the principal, predictive models for determining when chaos occurs. These include the period-doubling criterion, homoclinic orbit criterion, and the overlap criterion of Chirikov for conservative chaos, as well as intermittency and transient chaos. We also review several ad hoc criteria that have been developed for specific classes of problems.

Finally, in Section 5.4 we discuss an important diagnostic tool, namely, the Lyapunov exponent. Another diagnostic concept, the fractal dimension, is described in Chapter 6.

5.2 EMPIRICAL CRITERIA FOR CHAOS

In the many introductory lectures the author has given on chaos, the following question has surfaced time and time again: *Are chaotic motions singular cases in real physical problems or do they occur for a wide range of parameters?* For engineers this question is very important. To design, one needs to predict system behavior. If the engineer chooses parameters that produce chaotic output, then he or she loses predictability. In the past, many designs in structural engineering, electrical circuits, and control systems were kept within the realm of linear system dynamics. However, the needs of modern technology have pushed devices into nonlinear regimes (e.g., large deformations and deflections in structural mechanics) that increase the possibility of encountering chaotic dynamic phenomena.

To address the opening question, *are chaotic dynamics singular events in real systems*, we examine the range of parameters for which chaos occurs in seven different problems. A cursory scan of the figures accompanying each discussion will lead one to the conclusion that chaotic dynamics is not a singular class of motions and that *chaotic oscillations occur in many nonlinear systems for a wide range of parameter values.*

We examine the critical parameters for chaos in the following problems:

(a) Circuit with nonlinear inductor: Duffing's equation
(b) Particle in a two-well potential or buckling of an elastic beam: Duffing's equation
(c) Low-dimensional model for convection turbulence: Lorenz equations
(d) Vibrations of nonlinear coupled oscillators
(e) Rotating magnetic dipole: pendulum equation
(f) Circuit with nonlinear capacitance
(g) Surface waves on a fluid

(a) Forced Oscillations of a Nonlinear Inductor: Duffing's Equation

In Chapter 3 (Figure 3-13), we examined the chaotic dynamics of a circuit with a nonlinear inductor. Extensive analog and digital simulation for this system was performed by Y. Ueda (1979, 1980) of Kyoto University. The nondimensional equation, where x represents the flux in the inductor, takes the form

$$\ddot{x} + k\dot{x} + x^3 = B\cos t \qquad (5\text{-}2.1)$$

The time has been nondimensionalized by the forcing frequency so that the entire dynamics is determined by the two parameters k and B and the initial conditions $(x(0), \dot{x}(0))$. Here k is a measure of the resistance of the circuit, while B is a measure of the driving voltage. Ueda found that by varying these two parameters one could obtain a wide variety of periodic, subharmonic, ultrasubharmonic as well as chaotic motions. The regions of chaotic behavior in the (k, B) plane are plotted in Figure 5-1. The regions of subharmonic and harmonic motions are quite complex and only a few are shown for illustration. The two different hatched areas indicate either regions of only chaos, or regions with both chaotic as well as periodic

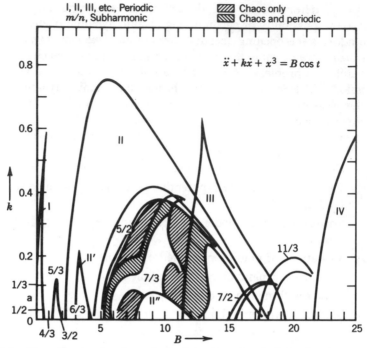

Figure 5-1 Chaos diagram showing regions of chaotic and periodic motions for a nonlinear circuit as functions of nondimensionalized damping and forcing amplitude [from Ueda (1980)].

motion depending on initial conditions. A theoretical criterion for this relatively simple equation has not been found to date.

(b)   Forced Oscillations of a Particle in a Two-Well Potential: Duffing's Equation

This example was discussed in great detail in Chapters 2 and 3. It was first studied by Holmes (1979) and later in a series of papers by the author and coworkers. The mathematical equation describes the forced motion of a particle between two states of equilibrium, which can be described by a two-well potential

$$\ddot{x} + \delta\dot{x} - \tfrac{1}{2}x(1 - x^2) = f\cos\omega t \qquad (5\text{-}2.2)$$

This equation can represent a particle in a plasma, a defect in a solid, and, on a larger scale, the dynamics of a buckled elastic beam (see Chapter 3). The dynamics are controlled by three nondimensional groups (δ, f, ω),

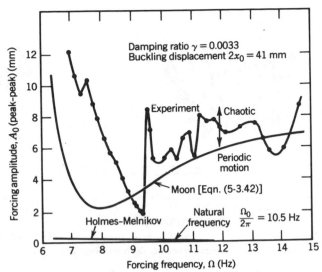

Figure 5-2 Experimental chaos diagram for vibrations of a buckled beam for different values of forcing frequency and amplitude [from Moon (1980b); reprinted with permission from *New Approaches to Nonlinear Problems in Dynamics*, ed. by P. S. Holmes; copyright 1980 by SIAM].

where δ represents nondimensional damping and ω is the driving frequency nondimensionalized by the small amplitude natural frequency of the system in one of the potential wells.

Regions of chaos from two studies are shown in Figures 5-2 and 5-3. The first represents experimental data for a buckled cantilevered beam (see Chapter 2). The ragged boundary is the experimental data, while the smooth curve represents a theoretical criterion (see Section 5.3). Recently, an upper boundary has been measured beyond which the motion again becomes periodic. The experimental criterion was determined by looking at Poincaré maps of the motion (see Chapters 2 and 4).

Results from numerical simulation of Eq. (5-2.2) are shown in Figure 5-3. The diagnostic tool used to determine if chaos were present was the Lyapunov exponent using a computer algorithm developed by Wolf et al. (1985) (see Section 5.4). This diagram shows that there are complex regions of chaotic vibrations in the plane (f, ω) for fixed damping δ. For very large forcing $f \gg 1$, one expects the behavior to emulate the previous problem studied by Ueda.

The theoretical boundary found by Holmes (1979) is discussed in the next section. It has special significance since below this boundary periodic motions are predictable, while above this boundary one loses the ability to exactly predict to which of the many periodic or chaotic modes the motion

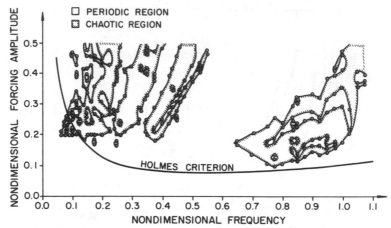

Figure 5-3 Chaos diagram for vibration of a mass in a double-well potential [Duffing's equation (5-2.2)]. The smooth boundary represents the homoclinic orbit criterion (Section 5.3).

will be attracted. Above the theoretical criteria (based on homoclinic orbits), the motion is very sensitive to initial conditions, even when it is periodic.

(c)   Rayleigh–Benard Convection: Lorenz Equations

Aside from the logistic equation, the Lorenz model for convection turbulence (see Chapters 1 and 3) is perhaps the most studied system of equations that admit chaotic solutions. Yet most mathematicians have focused on a few sets of parameters. These equations take the form

$$\dot{x} = \sigma(y - x)$$

$$\dot{y} = rx - y - xz \qquad\qquad (5\text{-}2.3)$$

$$\dot{z} = xy - bz$$

Sparrow (1982) in his book on the Lorenz attractor concentrates his analysis on the parameter values $\sigma = 10$, $b = \frac{8}{3}$, $r > 14$. However, he does speculate on the range of values for which stable chaotic motions might exist as reproduced in Figure 5-2. The vertical hatched region in Figure 5-4 represents a region of steady chaotic motion, while the horizontally hatched region represents a preturbulent region in which there may be chaotic transients. This region is bounded below by a criterion based on the

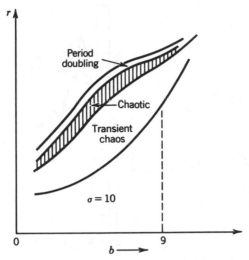

Figure 5-4   Chaos diagram for Lorenz's equations (5-2.3) for thermal convection dynamics.

existence of homoclinic orbits (see next section). A period-doubling region is also shown in the dotted region.

(d)   Forced Vibrations of a Two-Degree-of-Freedom Oscillator in a Two-Well Potential

As extension of the one-degree-of-freedom particle in a two-well potential has been studied by the author for the experiment shown in Figure 5-5. This problem can be modeled by two coupled nonlinear oscillators (3-3.7),

$$\ddot{x} + \gamma\dot{x} + \frac{\partial V}{\partial x} = 0$$

$$\ddot{y} + \gamma\dot{y} + \frac{\partial V}{\partial y} = f\cos\omega t$$

(5-2.4)

where $V(x, y)$ represents the potential for the magnets and the elastic stiffness. The chaotic regime for the forcing amplitude and frequency are shown in Figure 5-5. Comparing this regime to that in either Figure 5-2 or 5-3, we see that the addition of the extra degree of freedom seems to have reduced the extent of the chaos region at least in the vicinity of the natural frequency of the mass in one of the potential wells.

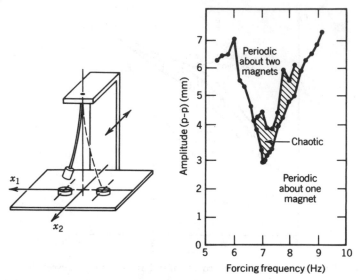

Figure 5-5 Regions of chaotic and periodic motions for two-dimensional motion of a mass in a double-well potential in the forcing amplitude–frequency plane [from Moon (1980b); reprinted with permission from *New Approaches to Nonlinear Problems* in *Dynamics*, ed. by P. S. Holmes; copyright 1980 by SIAM].

(e)   Forced Motions of a Rotating Dipole in Magnetic Fields: The Pendulum Equation

In this experiment, a permanent magnet rotor is excited by crossed steady and time harmonic magnetic fields (see Moon et al., 1987), as shown in Figure 3-18. The nondimensionalized equation of motion for the rotation angle θ resembles that for the pendulum in a gravitational potential:

$$\ddot{\theta} + \gamma\dot{\theta} + \sin\theta = f\cos\theta\cos\omega t \qquad (5\text{-}2.5)$$

The regions of chaotic rotation in the f–ω plane, for fixed damping, are shown in Figure 5-6. This is one of the few published examples where both experimental and numerical simulation data are compared with a theoretical criterion for chaos. The theory is based on the homoclinic orbit criterion and is discussed in Section 5.4. As in the case of the two-well potential, chaotic motions are to be found in the vicinity of the natural frequency for small oscillations ($\omega = 1.0$ in Figure 5-6).

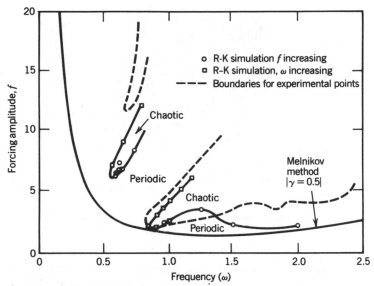

Figure 5-6 Experimental chaos diagram for forced motions of a rotor with nonlinear torque–angle property. Comparison with homoclinic orbit criterion calculated using the Melnikov method (Section 5.3) [From Moon et al. (1987) with permission of North-Holland Publishing Co., copyright 1987].

(f) Forced Oscillations of a Nonlinear *RLC* Circuit

There have been a number of experimental studies of chaotic oscillations in nonlinear circuits (e.g., see Chapter 3). One example is a *RLC* circuit with a diode. Shown in Figure 5-7 are the subharmonic and chaotic regimes in the driving voltage–frequency plane (Klinker et al., 1984). In this example, regions of period doubling are shown as precursors to the chaotic motions. However, in the midst of the hatched chaotic regime, a period 5 subharmonic was observed. Periodic islands in the center of chaotic domains are common observations in experiments on chaotic oscillations. (See a similar study by Bucko et al., 1984. See also Figure 3-33.).

(g) Harmonically Driven Surface Waves in a Fluid Cylinder

As a final example, we present experimentally determined harmonic and chaotic regions of the amplitude–frequency parameter space for surface waves in a cylinder filled with water from a paper by Cilberto and Gollub (1985). A 12.7 cm diameter cylinder with 1 cm deep water was harmonically

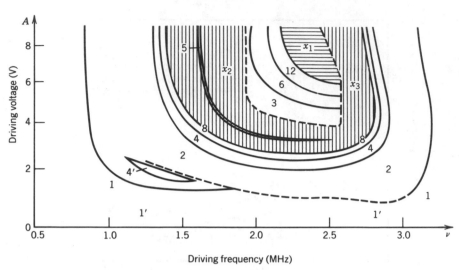

Driving frequency (MHz)

Figure 5-7 Experimentally determined chaos diagram for a driven *RLC* circuit with a varactor diode that acts as a nonlinear capacitor. The hatched regions are chaotic motions while the numbers indicate the order of the subharmonic. Dashed lines indicate a hysteretic transition [from Klinker et al. (1984) with permission of North-Holland Publishing Co., copyright 1984].

vibrated in a speaker cone (Figure 5-8). The amplitude of the transverse vibration above the flat surface of the fluid can be written in terms of Bessel functions where the linear mode shapes are given by

$$U_{nm} = J_n(k_{nm}r)\sin(n\theta + d_{nm})$$

Figure 5-8 shows the driving amplitude–frequency plane in a region where two modes can interact—$(n, m) = (4, 3)$ and $(7, 2)$. Below the lower boundary, the surface remains flat. A small region of chaotic regimes intersect. Presumably, other chaotic regimes exist where other modes (n, m) interact.

In summary, these examples show that, given periodic forcing input to a physical system, large regions of periodic or subharmonic motions do exist and presumably are predictable using classical methods of nonlinear analysis. However, these examples also show that *chaos is not a singular happening*; that is, it can exist for wide ranges in the parameters of the problem. Also, and perhaps this is most important, there are regions where both

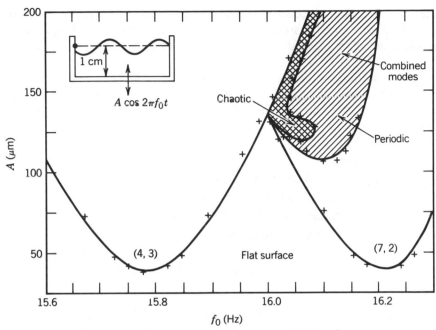

Figure 5-8 Experimental chaos diagram for surface waves in a cylinder filled with water. The diagram shows where two linear modes interact [from Ciliberto and Gollub (1985)].

periodic and chaotic motions can exist and the precise motion that will result may be unpredictable.

5.3   THEORETICAL PREDICTIVE CRITERIA

The search for theoretical criteria to determine under what set of conditions a given dynamical system will become chaotic has tended to be ad hoc. The strategy thus far has been for theorists to find criteria for specific mathematical models and then use these models as analogs or paradigms to infer when more general or complex physical systems will become unpredictable. An example is the period-doubling bifurcation sequence discussed by May (1976) and Feigenbaum (1978) for the quadratic map (e.g., see Chapter 1). Although these results were generalized for a wider class of one-dimensional maps using a technique called renormalization theory, the period-doubling criterion is not always observed for higher-dimensional maps. In mechanical and electrical vibrations, a Poincaré section of the solution in phase space often leads to maps of two or higher dimensions. Nonetheless, the period-doubling scenario is one possible route to chaos. In more complicated physical systems, an understanding of the May–Feigenbaum model can be very useful in determining when and why chaotic motions occur.

In this section, we briefly review a few of the principal theories of chaos and explore how they lead to criteria that may be used to predict or diagnose chaotic behavior in real systems. These theories include the following:

(a) Period doubling
(b) Homoclinic orbits and horseshoe maps
(c) Intermittency and transient chaos
(d) Overlap criteria for conservative chaos
(e) Ad hoc theories for multiwell potential problems

(a) Period-Doubling Criterion

This criterion is applicable to dynamical systems whose behavior can be described exactly or approximately by a first-order difference equation, known in the new jargon as a one-dimensional map:

$$x_{n+1} = \lambda x_n (1 - x_n) . \tag{5-3.1}$$

The dynamics of this equation were studied by May (1976), Feigenbaum (1978, 1980), and others. They discovered solutions whose period doubles as the parameter λ is varied (the period in this case is the number of integers p for x_{n+p} to return to the value x_n). One of the important properties of Eq. (5-3.1) that Feigenbaum discovered was that the sequence of critical parameters $\{\lambda_m\}$ at which the period of the orbit doubles satisfies the relation

$$\lim_{m \to \infty} \frac{\lambda_{m+1} - \lambda_m}{\lambda_m - \lambda_{m-1}} = \frac{1}{\delta}, \qquad \delta = 4.6692 \cdots \tag{5-3.2}$$

This important discovery gave experimenters a specific criterion to determine if a system was about to become chaotic by simply observing the prechaotic periodic behavior. It has been applied to physical systems involving fluid, electrical, and laser experiments. Although these problems are often modeled mathematically by continuous differential equations, the Poincaré map can reduce the dynamics to a set of difference equations. For

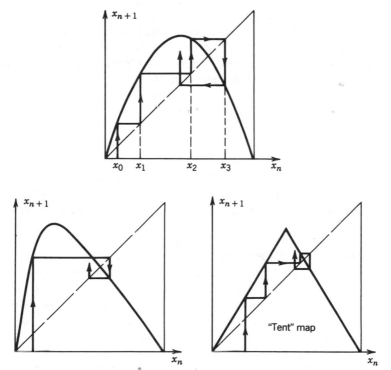

Figure 5-9 One hump noninvertible difference equations (maps) which exhibit period doubling.

many physical problems, the essential dynamics can be modeled further as a one-dimensional map.

$$x_{n+1} = f(x_n) \tag{5-3.3}$$

The importance of Feigenbaum's work is that he showed how period-doubling behavior was typical of one-dimensional maps that have a hump or zero tangent as shown in Figure 5-9 [i.e., the map is *noninvertible* or there exists two values of x_n which when put into $f(x_n)$ give the same value of x_{n+1}]. He also demonstrated that if the mapping function depends on some parameter Λ, that is, $f(x_n; \Lambda)$, then the sequence of critical values of this parameter at which the orbit's period doubles $\{\Lambda_m\}$ satisfies the same relation (5-3.2) as that for the quadratic map. Thus, the period-doubling phenomenon has been called *universal* and δ has been called a universal constant (now known quite naturally as the *Feigenbaum number*).

The author must raise a flag of caution here. The term "universal" is used in the context of one-dimensional maps (5-3.3). There are many chaotic phenomena which are described by two- or higher-dimensional maps (e.g., see the buckled beam problem in Chapter 2). In these cases, period doubling may indeed be one route to chaos, but there are many other bifurcation sequences that result in chaos beside period doubling (see Holmes, 1984).

The reader wishing a very detailed mathematical discussion of the quadratic map and period doubling should see either Lichtenberg and Lieberman (1983) or Guckenheimer and Holmes (1983). For the reader who desires a taste of the mathematics of period doubling, a distilled version of the treatment by Lichtenberg and Lieberman, especially as it pertains to the critical parameter λ_∞ at which the motion becomes chaotic, is presented next.

Renormalization and the Period-Doubling Criterion. There are two ideas that are important in understanding the period-doubling phenomenon. The first is the concept of *bifurcation* of solutions, and the second is the idea of *renormalization*. The concept of bifurcation is illustrated in Figure 5-10. The term bifurcate is used to denote the case where the qualitative behavior of the system suddenly changes as some parameter is varied. For example, in Figure 5-12 a steady periodic solution x_0 becomes unstable at a parameter value of λ, and the amplitude now oscillates between two values x_2^+ and x_2^-, completing a cycle in twice the time of the previous solution. Further changes in λ make x_2^+ and x_2^- unstable and the solution branches to a new cycle with period 4. In the case of the quadratic map (5-3.1), these solution bifurcations continue ad infinitum as λ is increased (or decreased). However, the critical values of λ approach an accumulation value, that is, $\lim_{i \to \infty} |\lambda_i| = |\lambda_\infty| < \infty$, beyond which the system can exhibit a chaotic, nonperiodic solution. Thus, if λ is some nondimensional function of physical variables (e.g., a Reynolds number in fluid mechanics), $\lambda = \lambda_\infty$ becomes a useful criterion to predict when chaos is likely to occur.

A readable description of renormalization as it applies to period doubling may be found in Feigenbaum (1980). The technique recognizes the fact that a cascade of bifurcations exists and that it might be possible to map each bifurcation into the previous one by a change in scale of the physical variable x and a transformation of the control parameter. To illustrate this technique, we outline an approximate scheme for the quadratic map (see also Lichtenberg and Lieberman, 1983).

One form of the quadratic map is given by

$$x_{n+1} = f(x_n)$$

where $f(x) = \lambda x(1 - x)$. Period 1 cycles are just constant values of x given

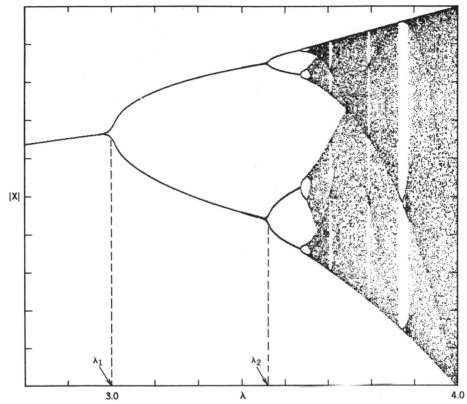

Figure 5-10 Bifurcation diagram for the quadratic map (5-3.3). Steady-state behavior as a function of the control parameter showing period-doubling phenomenon.

by fixed points of the mapping, that is, $x_n = f(x_n)$, or

$$x_0 = \lambda x_0 (1 - x_0) \qquad (5\text{-}3.4)$$

which gives $x_0 = 0$, $x_0 = (\lambda - 1)/\lambda$. Now a fixed point, or equilibrium point can be stable or unstable. That is, iteration of x can move toward or away from x_0. The stability of the map depends on the slope of $f(x)$ at x_0; that is,

$$\left| \frac{df(x_0)}{dx} \right| < 1 \quad \text{implies stability}$$

$$\left| \frac{df(x_0)}{dx} \right| > 1 \quad \text{implies instability}$$

$$(5\text{-}3.5)$$

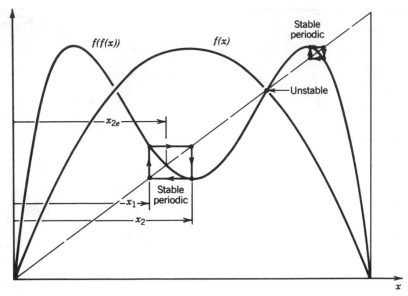

Figure 5-11　First and second iteration functions for the quadratic map (5-3.3) [see also Eq. (5-3.6)].

Since the slope $f' = \lambda(1 - 2x)$ depends on λ, x_0 becomes unstable at $\lambda_1 = \pm 1/|1 - 2x_0|$. Beyond this value, the stable periodic motion has period 2. The fixed points of the period 2 motion are given by

$$x_2 = f(f(x_2)) \quad \text{or} \quad x_2 = \lambda^2 x_2(1 - x_2)[1 - \lambda x_2(1 - x_2)] \quad (5\text{-}3.6)$$

The function $f(f(x))$ is shown in Figure 5-11.

Again there are stable and unstable solutions. Suppose the x_0 solution bifurcates and the solution alternates between x^+ and x^- as shown in Figure 5-12. We then have

$$x^+ = \lambda x^-(1 - x^-) \quad \text{and} \quad x^- = \lambda x^+(1 - x^+) \quad (5\text{-}3.7)$$

To determine the next critical value $\lambda = \lambda_2$ at which a period 4 orbit emerges, we change coordinates by writing

$$x_n = x^{\pm} + \eta_n \quad (5\text{-}3.8)$$

Putting Eq. (5-3.8) into (5-3.7), we get

$$\eta_{n+1} = \lambda \eta_n[(1 - 2x^+) - \eta_n]$$
$$\eta_{n+2} = \lambda \eta_{n+1}[(1 - 2x^-) - \eta_{n+1}] \quad (5\text{-}3.9)$$

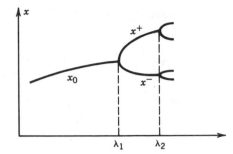

Figure 5-12 Diagram showing two branches of a bifurcation diagram near a period-doubling point.

We next solve for η_{n+2} in terms of η_n, keeping only terms to order η_n^2 (this is obviously an approximation), to obtain

$$\eta_{n+2} = \lambda^2 \eta_n [A - B\eta_n] \qquad (5\text{-}3.10)$$

where A and B depend on x^+, x^-, and λ. Next, we rescale η and define a new parameter $\bar{\lambda}$ using

$$\bar{x} = \alpha \eta, \qquad \bar{\lambda} = \lambda^2 A, \qquad \alpha = B/A$$

$$\bar{x}_{n+2} = \bar{\lambda}\bar{x}_n(1 - \bar{x}_n)$$

This has the same form as our original equation (5-3.2). Thus, when the solution bifurcates to period 4 at $\lambda = \lambda_2$, the critical value of $\bar{\lambda} = \lambda_1$. We therefore obtain an equation

$$\lambda_1 = \lambda_2^2 A(\lambda_2) \qquad (5\text{-}3.11)$$

Starting from the point $x_0 = 0$, there is a bifurcation sequence for $\lambda < 0$. For this case Lichtenberg and Lieberman show that (5-3.11) is given by

$$\lambda_1 = -\lambda_2^2 + 2\lambda_2 + 4 \qquad (5\text{-}3.12)$$

It can be shown that $\lambda_1 = -1$, so that $\lambda_2 = (1 - \sqrt{6}) = -1.4494$. If one is bold enough to propose that the recurrence relation (5-3.12) holds at higher-order bifurcations, then

$$\lambda_\kappa = -\lambda_{\kappa+1}^2 + 2\lambda_{\kappa+1} + 4 \qquad (5\text{-}3.13)$$

At the critical value for chaos,

$$\lambda_\infty = -\lambda_\infty^2 + 2\lambda_\infty + 4$$

$$= (1 - \sqrt{17})/2 = -1.562 \qquad (5\text{-}3.14)$$

One can also show that another bifurcation sequence occurs for $\lambda > 0$ (Figure 5-10) where the critical value is given by,

$$\lambda = \hat{\lambda}_\infty = 2 - \lambda_\infty = 3.56 \qquad (5\text{-}3.15)$$

The exact value is close to $\lambda_\infty = 3.56994$. Thus, the rescaling approximation scheme is not too bad.

This line of analysis also leads to the relation

$$\lambda_\kappa \simeq \lambda_\infty + a\delta^{-\kappa} \qquad (5\text{-}3.16)$$

which results in the scaling law (5-3.2). Thus, knowing that two successive bifurcation values can give one an estimate of the chaos criterion λ_∞, we obtain

$$\lambda_\infty \simeq \frac{1}{(\delta - 1)}[\delta\lambda_{\kappa+1} - \lambda_\kappa] \qquad (5\text{-}3.17)$$

A final word before we leave this section: The fact that λ may exceed the critical value ($|\lambda| > |\lambda_\infty|$) does not imply that chaotic solutions will occur. They certainly are possible. But there are also many *periodic windows* in the range of parameters greater than the critical value in which periodic motions as well as chaotic solutions can occur.

We do not have space to do complete justice to the rich complexities in the dynamics of the quadratic map. It is certainly one of the major paradigms for understanding chaos and the interested reader is encouraged to study this problem in the aforementioned references. (See also Appendix B for computer experiments.)

(b) Homoclinic Orbits and Horseshoe Maps

One theoretical technique that has led to specific criteria for chaotic vibrations is a method based on the search for horseshoe maps and homoclinic orbits in mathematical models of dynamical systems. This strategy and a mathematical technique, called the Melnikov method, has led to Reynolds-numberlike criteria for chaos relating the parameters in the system. In two cases, these criteria have been verified by numerical and physical experiments. Keeping with the tenor of this book, we do not derive or go into too much of the mathematical theory of this method, but we do try to convey the rationale behind it and guide the reader to the literature for a more detailed discussion of the method. We illustrate the Melnikov method with two applications: the vibrations of a buckled beam and the rotary dynamics of a magnetic dipole motor.

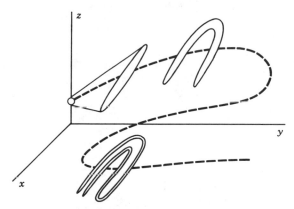

Figure 5-13   Evolution of an initial condition sphere.

The homoclinic orbit criterion is a mathematical technique for obtaining a predictive relation between the nondimensional groups in the physical system. It gives one a necessary but not sufficient condition for chaos. It may also give a necessary and sufficient condition for predictability in a dynamical system (see Chapter 6, Section 6.5, Fractal Basin Boundaries). Stripped of its complex, somewhat arcane mathematical infrastructure, it is essentially a method to prove whether a model in the form of partial or ordinary differential equations has the properties of a horseshoe or a baker's-type map.

The horseshoe map view of chaos (see also Chapter 1) looks at a collection of initial condition orbits in some ball in phase space. If a system has a horseshoe map behavior, this initial volume of phase space is mapped under the dynamics of the system onto a new shape in which the original ball is stretched and folded (Figure 5-13). After many iterations, this folding and stretching produces a fractal-like structure and the precise information as to which orbit originated where is lost. More and more precision is required to relate an initial condition to the state of the system at a later time. For a finite precision problem (as most numerical or laboratory experiments are), predictability is not possible.

One path to an understanding of the homoclinic orbit criterion (see the flow chart in Figure 5-14) is to go through the following questions:

1. What are homoclinic orbits?
2. How do homoclinic orbits arise in mathematical models of physical systems?
3. How are they related to horseshoe maps?
4. Finally, how does the Melnikov method lead to a criterion for chaos?

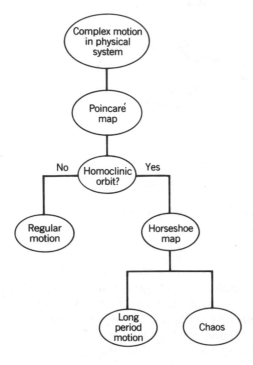

Figure 5-14 Diagram showing the relation between homoclinic orbits, horseshoes, and chaos in physical systems.

Homoclinic Orbits. A good discussion of homoclinic orbits may be found in the books by Lichtenberg and Lieberman (1983) and Guckenheimer and Holmes (1983). We have learned earlier that although many dynamics problems can be viewed as a continuous curve in some phase space (x versus $v = \dot{x}$) or solution space (x versus t), the mysteries of nonlinear dynamics and chaos are often deciphered by looking at a digital sampling of the motion such as a Poincaré map. We have also seen that the Poincaré map, although a sequence of points in some n-dimensional space, can lie along certain continuous curves. These curves are called *manifolds*. A discussion of homoclinic orbits refers to a sequence of points. This sequence of points is called an orbit. For example, for a period 3 orbit, the sequence of points will sequentially visit three states in the phase plane as in Figure 5-15a. On the other hand, a *quasiperiodic orbit* will involve a sequence of points that move on some closed curve, as in Figure 5-15b. Quasiperiodic vibrations are common in the motion of two coupled oscillators with two incommensurate frequencies.

In the dynamics of mappings, one can have critical points at which orbits move away from or toward. One example is a saddle point at which there are two manifold curves on which orbits approach the point and two curves

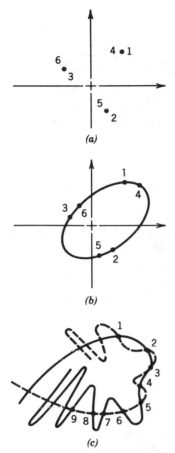

Figure 5-15 (*a*) Periodic orbit in a Poincaré map. (*b*) Quasiperiodic orbit. (*c*) Homoclinic orbit.

on which the sequence of Poincaré points move away from the point, as illustrated in Figure 5-15. Such a point is similar to a *saddle point* in nonlinear differential equations.

To illustrate a homoclinic orbit, we consider the dynamics of the forced damped pendulum. First, recall that for the unforced damped pendulum, the unstable branches of the saddle point swirl around the equilibrium point in a vortexlike motion in the θ–$\dot{\theta}$ phase plane as shown in Figure 5-16.

Although it is not obvious, the Poincaré map synchronized with the forcing frequency also has a saddle point in the neighborhood of $\theta = \pm n\pi$ (*n* odd), as shown in Figure 5-17 for the case of the forced pendulum. For small forcing, the stable and unstable branches of the saddle do not touch each other. However, as the force is increased, these two manifolds inter-

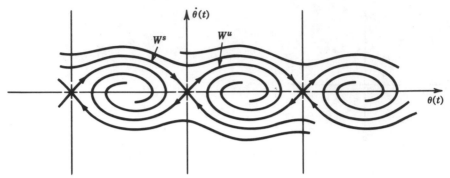

Figure 5-16 Stable and unstable manifolds for the motion of an unforced, damped pendulum.

sect. It can be shown that *if they intersect once*, they will *intersect an infinite number of times*. The points of intersection of stable and unstable manifolds are called *homoclinic points*. A Poincaré point near one of these points will be mapped into all the rest of the intersection points. This is called a *homoclinic orbit* (Figure 5-15c). Now why are these orbits important for chaos?

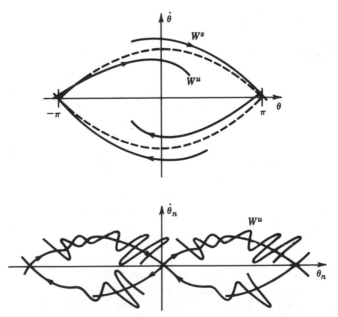

Figure 5-17 Sketch of stable and unstable manifolds of the Poincaré map for the harmonically forced, damped pendulum.

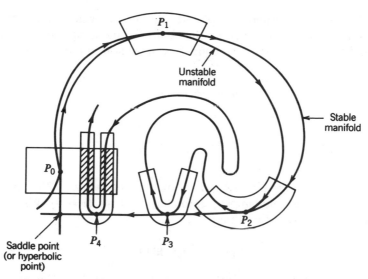

Figure 5-18 The development of a folded horseshoe-type map for points in the vicinity of a homoclinic orbit.

The intersection of the stable and unstable manifolds of the Poincaré map leads to a horseshoe-type map in the vicinity of each homoclinic point. As we saw in Chapter 1, horseshoe-type maps lead to unpredictability and unpredictability or sensitivity to initial conditions is a hallmark of chaos.

To see why homoclinic orbits lead to horseshoe maps, we recall that for a dissipative system areas get mapped into smaller areas. However, near the unstable manifold, the areas are also stretched. Since the total area must decrease, the area must also contract more than it stretches. Areas near the homoclinic points also get folded, as shown in Figure 5-18.

A dynamic process can be thought of as a transformation of phase space; that is, a volume of points representing different possible initial conditions is transformed into a distorted volume at a later time. Regular flow results when the transformed volume has a conventional shaped volume. Chaotic flows result when the volume is stretched, contracted, and folded as in the baker's transformation or *horseshoe* map.

The Melnikov Method. The Melnikov function is used to measure the distance between unstable and stable manifolds when that distance is small [see Guckenheimer and Holmes (1983) for a mathematical discussion of the Melnikov method]. It has been applied to problems where the dissipation is small and the equations for the manifolds of the zero dissipation problem are known. For example, suppose we consider the forced motion of a nonlinear oscillator where (q, p) are the generalized coordinate and momentum variables. We assume that both the damping and forcing are small and that we can write the equations of motion in the form

$$\dot{q} = \frac{\partial H}{\partial p} + \epsilon g_1$$

$$\dot{p} = -\frac{\partial H}{\partial q} + \epsilon g_2$$

(5-3.18)

where $\mathbf{g} = \mathbf{g}(p, q, t) = (g_1, g_2)$, ϵ is a small parameter, and $H(q, p)$ is the Hamiltonian for the undamped, unforced problems ($\epsilon = 0$). We also assume that $\mathbf{g}(t)$ is periodic so that

$$\mathbf{g}(t + T) = \mathbf{g}(t) \tag{5-3.19}$$

and that the motion takes place in a three-dimensional phase space $(q, p, \omega t)$, where ωt is the phase of the periodic force and is modulo the period T.

In many nonlinear problems a saddle point exists in the unperturbed Hamiltonian problem [$\epsilon = 0$ in Eq. (5-3.18)], such as for the pendulum or the double-well potential Duffing's equation, (5.22). When $\epsilon \neq 0$, one can take a Poincaré section of the three-dimensional torus flow synchronized with the phase ωt. It has been shown (see Guckenheimer and Holmes, 1983) that the Poincaré map also has a saddle point with stable and unstable manifolds, W^s and W^u, shown in Figure 5-19.

The Melnikov function provides a measure of the separation between W^s and W^u as a function of the phase of the Poincaré map ωt. This function is given by the integral

$$M(t_0) = \int_{-\infty}^{\infty} \mathbf{g}^* \cdot \nabla \mathbf{H}(q^*, p^*) \, dt \tag{5-3.20}$$

where $\mathbf{g}^* = \mathbf{g}(q^*, p^*, t + t_0)$ and $q^*(t)$ and $p^*(t)$ are the solutions for the unperturbed homoclinic orbit originating at the saddle point of the Hamiltonian problem. The variable t_0 is a measure of the distance along the original unperturbed homoclinic trajectory in the phase plane. We consider two examples.

Magnetic Pendulum. A convenient experimental model of a pendulum may be found in the rotary dynamics of a magnetic dipole in crossed steady and time periodic magnetic fields as shown in Figure 3-18 (See also Moon et al., 1987).

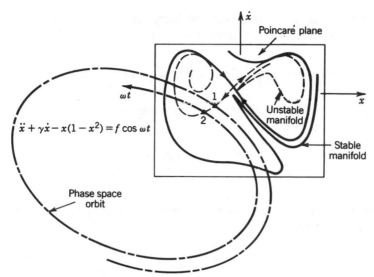

Figure 5-19 Saddle point of a Poincaré map and its associated stable and unstable manifolds before a homoclinic orbit develops.

The equation of motion when normalized is given by

$$\ddot{\theta} + \gamma\dot{\theta} + \sin\theta = f_1\cos\theta\cos\omega t + f_0 \qquad (5\text{-}3.21)$$

The $\sin\theta$ term is produced by the steady magnetic field, and the f_1 term is produced by the dynamic field. We have also included linear damping and a constant torque f_0. Keeping with the assumptions of the theory, we assume that one can write $\gamma = \epsilon\bar{\gamma}$, $f_0 = \epsilon\bar{f_0}$, and $f_1 = \epsilon\bar{f_1}$, where $0 \ll \epsilon < 1$ and $\bar{\gamma}$, $\bar{f_0}$, and $\bar{f_1}$ are of order one.

The Hamiltonian for the undamped, unforced problem is given by

$$H = \tfrac{1}{2}v^2 + (1 - \cos\theta)$$

where $q \equiv \theta$ and $p \equiv v = \dot{\theta}$. The energy H is constant $(H = 2)$ on the homoclinic orbit emanating from the saddle point $(\theta = v = 0)$. The unperturbed homoclinic orbit is given by

$$\theta^* = 2\tan^{-1}(\sinh t)$$

$$v^* = 2\,\mathrm{sech}\,t \qquad (5\text{-}3.22)$$

In Eq. (5-3.18), $g_1 = 0$ and $g_2 = f_0 + f_1\cos\theta\cos\omega t$. The resulting integral can be carried out exactly using contour integration (e.g., see Guckenheimer and Holmes for a similar example). The result gives

$$M(t_0) = -8\bar{\gamma} + 2\pi\bar{f}_0 + 2\pi\bar{f}_1\omega^2 \operatorname{sech}\left(\frac{\pi\omega}{2}\right)\cos \omega t_0 \qquad (5\text{-}3.23)$$

The two perturbed manifolds will touch transversely when $M(t_0)$ has a simple zero, or when

$$f_1 > \left|\frac{4\gamma}{\pi} - f_0\right|\frac{\cosh(\pi\omega/2)}{\omega^2} \qquad (5\text{-}3.24)$$

where we have canceled the ϵ factors. When $f_0 = 0$, the critical value of the forcing torque is given by

$$f_{1c} = \frac{4\gamma}{\pi\omega^2}\cosh\left(\frac{\pi\omega}{2}\right) \qquad (5\text{-}3.25)$$

This function is plotted in Figure 5-6 along with experimental and numerical simulation data. The criterion (5-3.25) gives a remarkably good lower bound on the regions of chaos in the forcing amplitude–frequency plane.

Two-Well Potential Problem. Forced motion of a particle in a two-well potential has numerous applications such as postbuckling behavior of a buckled elastic beam (Moon and Holmes, 1979) or certain plasma dynamics (Mahaffey, 1976). Damped periodically forced oscillations can be described by a Duffing-type equation

$$\ddot{x} + \gamma\dot{x} - x + x^3 = f\cos \omega t \qquad (5\text{-}3.26)$$

The Hamiltonian for the unperturbed problem is

$$H(x, v) = \tfrac{1}{2}\left(v^2 - x^2 + \tfrac{1}{2}x^4\right)$$

For $H = 0$, there are two homoclinic orbits originating and terminating at the saddle point at the origin. The variables x^* and v^* take on values along the right half plane curve given by

$$x^* = \sqrt{2}\operatorname{sech} t \quad \text{and} \quad v^* = -\sqrt{2}\operatorname{sech} t \tanh t$$

In this problem, $g_1 = 0$ and $g_2 = \bar{f}\cos \omega t - \bar{\gamma}v$, where $\gamma = \epsilon\bar{\gamma}$ and $f = \epsilon\bar{f}$ as in the previous example. The Melnikov function (5-3.20) then takes the form

$$M(t_0) = -\sqrt{2} f \int_{-\infty}^{\infty} \text{sech}\, t \tanh t \cos \omega (t + t_0)\, dt - 2\gamma \int_{-\infty}^{\infty} \text{sech}^2 t \tanh^2 t \, dt$$

which can be integrated exactly using methods of contour integration. The solution was originally found by Holmes (1979) but an error crept into his paper. The correct analysis is in Guckenheimer and Holmes (1983):

$$M(t_0) = -\frac{4\gamma}{3} - \sqrt{2} f \pi \omega \, \text{sech}\left(\frac{\pi\omega}{2}\right) \sin \omega t_0 \qquad (5\text{-}3.27)$$

For a simple zero we require

$$f > \frac{4\gamma}{3} \frac{\cosh(\pi\omega/2)}{\sqrt{2}\,\pi\omega} \qquad (5\text{-}3.28)$$

This lower bound on the chaotic region in (f, ω, γ) space has been verified in experiments by Moon (1980a) (see also Figures 5-2, 5-3).

(c) Intermittent and Transient Chaos

Thus far we have discussed what one might call "steady-state" chaotic vibration. Two other forms of unpredictable, irregular motions are intermittency and transient chaos. In the former, bursts of chaotic or noisy motion occur between periods of regular motion (see Figure 5-20). Such behavior was even observed by Reynolds in pipe flow preturbulence experiments in 1883 (see Sreenivasan, 1986). Transient chaos is also observed in some systems as a precursor to steady-state chaos. For certain initial conditions, the system may behave in a randomlike way, with the trajectory moving in phase space as if it were on a strange attractor. However, after some time,

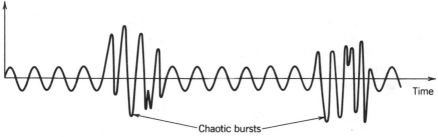

Time

Chaotic bursts

Figure 5-20   Sketch of intermittent chaotic motion.

the motion settles onto a regular attractor such as a periodic vibration. Scaling properties of nonlinear motion can sometimes be used to determine experimentally a critical parameter for these two types of chaotic motion. In the case of intermittency, where the dynamic system is close to a periodic motion but experiences short bursts of chaotic transients, an explanation of this behavior has been posited by Manneville and Pomeau (1980) in terms of one-dimensional maps or difference equations.

From numerical experiments on maps, the mean time duration of the periodic motion between chaotic bursts $\langle \tau \rangle$ has been found to be

$$\langle \tau \rangle \sim \frac{1}{|\lambda - \lambda_c|^{1/2}} \tag{5-3.29}$$

where λ is a control parameter (e.g., fluid velocity, forcing amplitude, or voltage) and λ_c is the value at which a chaotic motion occurs. As $\lambda - \lambda_c$ increases, the chaotic time interval increases and the periodic interval decreases. Thus, one might call this *creeping chaos*.

To measure λ_c experimentally, one must measure two average times $\langle \tau \rangle_1$ and $\langle \tau \rangle_2$ at corresponding values of the control parameter, that is, λ_1 and λ_2. This should determine the proportionality constant in Eq. (5-3.29) as well as λ_c. Having obtained a candidate value for λ_c, however, one should then measure other values of $(\langle \tau \rangle, \lambda)$ to validate the scaling relation (5-3.29).

The case of transient chaos has been studied by Grebogi et al. (1983a, b, 1985b) of the University of Maryland in a series of papers describing numerical experiments on two-dimensional maps. In one study (1983), they investigate a two-dimensional extension of the one-dimensional quadratic difference equation called the Henon map (see also Section 1.3):

$$x_{n+1} = 1 - \alpha x_n^2 + y_n$$

$$y_{n+1} = -Jx_n$$

J is the determinant of the Jacobian matrix which controls the amount of area contraction of the map. In the Maryland group's research on transient chaos, the case of $J = -0.3$ with the parameter α varied was investigated. For example, for $\alpha > \alpha_0 = 1.062371838$, a period-6 orbit gave birth to a six-piece strange attractor that exists in the region

$$\alpha_0 < \alpha < \alpha_c = 1.080744879$$

For $\alpha > \alpha_c$, the orbit under the iteration of the Henon map was found to wander around the ghost of the strange attractor in the x–y plane, sometimes for over 10^3 iterations, before settling onto a period-4 motion.

They also discovered that the average time for the transient chaos $\langle \tau \rangle$ followed a scaling law

$$\langle \tau \rangle \sim (\alpha - \alpha_c)^{-1/2} \qquad (5\text{-}3.30)$$

The average was found by choosing 10^2 initial conditions for each choice of α. The initial conditions were chosen in the original basin of attraction of the defunct strange attractor. These transients can be very long. For example, in the case of the Henon map, Grebogi and coworkers found $\langle \tau \rangle \approx 10^4$ for $\alpha - \alpha_c = 5 \times 10^{-7}$ and $\langle \tau \rangle \approx 10^3$ for $\alpha - \alpha_c = 10^{-5}$.

This same research group has also found maps that exhibit *supertransient chaos* in which the transient lifetime scales as (see Grebogi et al., 1985b)

$$\langle \tau \rangle \geq k_1 \exp\left[k_2 (\alpha - \alpha_c)^{-1/2} \right] \qquad (5\text{-}3.31)$$

These results suggest that some transients could live beyond the practical life of any experiment. The mathematics relating to these studies again involves homoclinic intersections of stable and unstable manifolds in maps. The Maryland group refers to such homoclinic tangencies as *crises*. A full discussion of the mathematics concerning transient chaos is beyond the scope of this book and the reader is referred to the original work of the Maryland group cited above.

Unfortunately, few if any physical examples or experiments have been associated with the study of transient chaos thus far. But it would appear to be a fertile subject for further study.

(d) Chirikov's Overlap Criterion for Conservative Chaos

The study of chaotic motions in conservative systems (no damping) predates the current interest in chaotic dissipative systems. Since the practical application of conservative dynamical systems is limited to areas such as planetary mechanics, plasma physics, and accelerator physics, engineers have not followed this field as closely as other advances in nonlinear dynamics.

In this section, we focus on the bouncing ball chaos described in Chapter 3 (Figure 3-5). However, the resulting difference equations are relevant to the behavior of coupled nonlinear oscillators (e.g., see Lichtenberg and Lieberman, 1983) as well as the behavior of electrons in electromagnetic fields. The equations for the impact of a mass, under gravity, on a vibrating

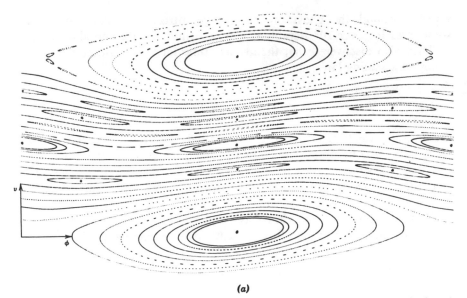

(a)

Figure 5-21 (a) Poincaré map for elastic motion of a ball on a vibrating table (standard map) for the parameter $K = 0.6$ in Eq. (5-3.32) showing periodic and quasiperiodic orbits.

table are given by (3-2.9) with a change of variables these become

$$v_{n+1} = v_n + K \sin \varphi_n$$
$$\varphi_{n+1} = \varphi_n + v_{n+1}$$

(5-3.32)

where v_n is the velocity before impact and φ_n is the time of impact normalized by the frequency of the table (i.e., $\varphi = \omega t$ modulo 2π). K is proportional to the amplitude of the vibrating table in Figure 3-5. These equations differ from those in (3-2.9) by the assumption that there is no energy loss on impact. This implies that regions of initial conditions in the phase space (v, φ) preserve their areas under multiple iteration of the map (5-3.32).

Orbits in the (v, φ) plane for different initial conditions are shown in Figure 5-21 for two different values of K.

Consider the case of $K = 0.6$. The dots at $v = 0$, 2π correspond to period 1 orbits; that is,

$$v_1 = v_1 + K \sin \varphi_1$$
$$\varphi_1 = \varphi_1 + v_1$$

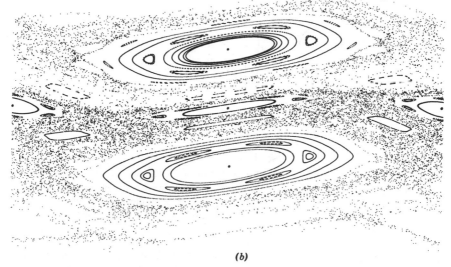

(b)

Figure 5-21   (*b*) The case of $K = 1.2$ showing the appearance of stochastic orbits.

whose solution is given by $\varphi_1 = 0, \pi, v_1 = 0$ (both mod 2π). The solution near $\varphi = \pi$ is stable for $|2 - K| < 2$. The solution near $\varphi = 0, 2\pi$, however, can be shown to be unstable for $|2 + K| < 2$ and can represent saddle points of the map.

Near $v = \pi$ one can see a period 2 orbit given by the solution to

$$v_2 = v_1 + K \sin \varphi_1, \qquad \varphi_2 = \varphi_1 + v_2$$

$$v_1 = v_2 + K \sin \varphi_2, \qquad \varphi_1 = \varphi_2 + v_1$$

Again one can show that there are both stable and unstable period 2 points. One can also show that the stable points exists as long as $K < 2$.

The rest of the continuous looking orbits in Figure 5-21 represent quasiperiodic solutions where the ball impact frequency is incommensurate with the driving period. Finally, a third type of motion is present in Figure 5-21 ($K = 1.2$). Here we see a diffuse set of dots near where saddle points and the saddle separatrices used to exist. This diffuse set of points represents *conservative chaos*. For $K < 1$, it is localized around the saddle points. However, for $K \approx 1$, this wandering orbit becomes global in nature.

The reader should note that in Figure 5-21 ($K = 0.6$) one can obtain all types of motion by simply choosing different initial conditions (since there is no damping, there are no attractors).

A criterion for global chaos in this system was proposed by the Soviet

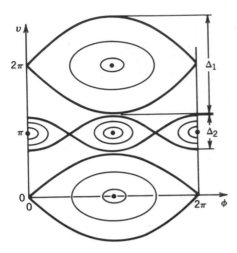

Figure 5-22 Sketch of period-1 and period-2 orbits and concomitant quasiperiodic orbits for the standard map used in the derivation of Chirikov's criterion.

physicist Chirikov (1979). He observed that as K is increased, the vertical distance between the separatrices associated with both period-1 and period-2 motion decreased. If chaos did not intervene, these separatrices would overlap (Figure 5-22)—thus the name *overlap criterion*.

If one performs a small-K analysis of the standard map (5-3.32) near one of these periodic resonances, the size of each separatrix region is found to be

$$\Delta_1 = 4K^{1/2}$$

$$\Delta_2 = K \tag{5-3.33}$$

Each analysis ignores the effect of the other resonance. The condition for overlap is that $\Delta_1 + \Delta_2 = 2\pi$, or

$$4K_c^{1/2} + K_c = 2\pi \tag{5-3.34}$$

The solution to this equation is $K_c = 1.46$. This value overestimates the critical value of $K = K_c$ for global chaos which is found numerically to be around $K_c \approx 1.0$. The reader is referred to Lichtenberg and Lieberman (1983) for further details concerning the overlap criterion.

The more practical minded reader might ask: *What happens when we have a small amount of damping present?* For that case, some of the multiperiod subharmonics become attractors and the ellipses surrounding these attractors become spirals that limit the periodic motions. *What of the conservative chaos?* Initial conditions in regions where there was conservative chaos become long chaotic transients which wander around phase space before settling into a periodic motion. *And what about real chaotic motions*? When damping is present, one needs a much larger force, $K > 6$, for which a fractal-like strange attractor appears (see Figure 3-5). Thus, the overlap criterion discussed above is only useful for strictly conservative, Hamiltonian systems.

(e) Multiwell Potential Criteria

In this section, we describe an ad hoc criterion for chaotic oscillations in problems with multiple potential energy wells. Such problems include the buckled beam (Chapter 2) and a magnetic dipole motor with multiple poles. In solid-state physics, interstitial atoms in a regular lattice can have more than one equilibrium position. Often the forces that create such problems can be derived from a potential. Let $\{q_i\}$ be a set of generalized coordinates and $V(q_i)$ be the potential associated with the conservative part of the force such that $-\partial V/\partial q_i$ is the generalized force associated with the q_i degree of freedom. For one degree of freedom, a special case might have the following equation of motion:

$$\ddot{q} + \gamma\dot{q} + \frac{\partial V}{\partial q} = f\cos\omega t \qquad (5\text{-}3.35)$$

where linear damping and periodic forcing have been added. $V(q_i)$ has as many local minima as stable equilibrium positions, as shown in Figure 5-23.

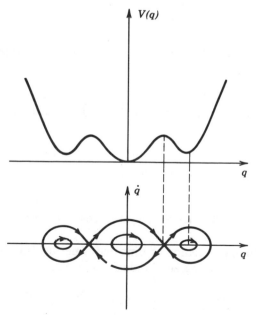

Figure 5-23 Multiwell potential energy function and associated phase plane.

For small periodic forcing, the system oscillates periodically in one potential well. But for larger forcing, the motion "spills over" into other wells and chaos often results. This criterion then seeks to determine *what value of the forcing amplitude will cause the periodic motion in one well to jump into another well.*

To illustrate the method, consider the particle in a two-well symmetric potential (i.e., the buckled beam problem of Chapter 2):

$$\ddot{q} + \gamma\dot{q} - \tfrac{1}{2}q(1 - q^2) = f\cos\omega t \qquad (5\text{-}3.36)$$

Since we are seeking a criterion that governs the transition from periodic to chaotic motion, we use standard perturbation theory to find a relation between the amplitude of forced motion $\langle q^2 \rangle$ (where $\langle\ \rangle$ indicates a time average) and the parameters γ, f, and ω. We then try to find a *critical* value of $\langle q^2 \rangle \equiv A_c$ independent of the forcing amplitude; that is,

$$\langle q^2 \rangle = g(\gamma, \omega, f) = A_c(\omega) \qquad (5\text{-}3.37)$$

The left-hand equality in Eq. (5-3.37) is found using classical perturbation theory, while the right-hand equality is based on a heuristic postulate.

To carry out this program for the two-well potential, we must write Eq. (5-3.36) in coordinates centered about one of the equilibrium positions:

$$\eta = q - 1$$

To obtain a perturbation parameter, we write $\eta = \mu X$, so that the equation of motion takes the form

$$\ddot{X} + \gamma\dot{X} + X\left(1 + \tfrac{3}{2}\mu X + \tfrac{1}{2}\mu^2 X^2\right) = \frac{f}{\mu}\cos(\omega t + \phi_0) \quad (5\text{-}3.38)$$

The phase angle ϕ_0 is adjusted so that the first-order motion is proportional to $\cos\omega t$. The resulting periodic motion for small f is assumed to take the form

$$X = C_0\cos\omega t + \mu(C_1 + C_2\cos\omega t) + \mu^2 X_1(t) \qquad (5\text{-}3.39)$$

Using either Duffing's method or Lindstedt's perturbation method (e.g., see Stoker, 1950), the resulting amplitude force relation can be found to be

$$(\mu C_0)^2\left\{\left[(1 - \omega^2) - \tfrac{3}{2}(\mu C_0)^2\right]^2 + \gamma^2\omega^2\right\} = f^2 \qquad (5\text{-}3.40)$$

Based on numerical experiments, we postulate the existence of a *critical velocity*. We propose that chaos is imminent when the maximum velocity of the motion is near the maximum velocity on the separatrix for the phase plane of the undamped, unforced oscillator. In terms of the original variables, this criterion becomes (see Figure 5-24)

$$\mu C_0 = \frac{\alpha}{2\omega} \qquad (5\text{-}3.41)$$

where α is close to unity. Substituting Eq. (5-3.41) into (5-3.40), we obtain a lower bound on the criterion for chaotic oscillations:

$$f_c = \frac{\alpha}{2\omega} \left\{ \left[(1 - \omega^2) - \frac{3\alpha^2}{8\omega^2} \right]^2 + \gamma^2 \omega^2 \right\}^{1/2} \qquad (5\text{-}3.42)$$

This expression has been checked against experiments by the author (Moon, 1980a) and a factor of $\alpha \approx 0.86$ seemed to give excellent agreement with experimental chaos boundaries as shown in an earlier figure (Figure 5-2). For low damping, this criterion gives a much better bound than the homoclinic orbit criterion using the Melnikov function.

As illustrated in Figure 5-24, this criterion is similar to the Chirikov overlap criterion—namely, that chaos results when a regular motion becomes too large.

The method outlined in this section has also been used on a three-well potential problem and has been tested successfully in experiments on a vibrating beam with three equilibria by Li (1985).

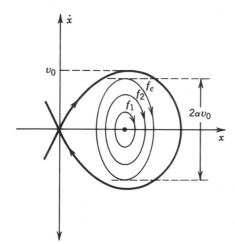

Figure 5-24 Overlap criteria for a multiwell problem using semiclassical analytic methods.

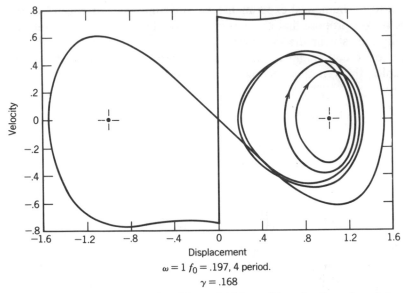

$$\omega = 1 \; f_0 = .197, \; 4 \; \text{period}.$$
$$\gamma = .168$$

Figure 5-25 Basins of attraction for different initial conditions for the unforced, two-well potential oscillator [from Dowell and Pezeshki (1986) with permission of the American Society of Mechanical Engineers, copyright 1985].

Dowell and Pezeshki (1986) have posited another heuristic criterion for the two-well potential problem (5-3.36). Instead of comparing the size of periodic orbits with the undamped, unforced problem, they compare the prechaotic, periodic, subharmonic orbits for the driven oscillator with the boundary of the basin of attraction for the *damped*, unforced problem (Figure 5-25). This boundary represents the set of initial conditions $(A(0), \dot{A}(0))$ for which the orbit goes to the left or right equilibrium point without crossing $A = 0$. They also observe, through numerical simulation, that the driven motion becomes chaotic when the force level of f_0 is larger than the value for which a periodic orbit touches the basin boundary. (See Chapter 6 for a discussion of basin boundaries.)

Criteria Derived from Classical Perturbation Analysis.   The novitiate to the field of nonlinear dynamics may be misled by the current interest in chaos to conclude that the field lay dormant in the prechaos era. However, a large literature exists describing mathematical perturbation methods for calculating primary and subharmonic resonances, as well as the stability characteristics of solutions to nonlinear systems (e.g., see Nehfeh and Mook, 1979). Thus, it is no surprise that studies are beginning to emerge that attempt to use the more classical analyses in the effort to find criteria for chaotic motion. For example, Nayfeh and Khdeir (1986) use perturbation techniques to predict the occurrence of period-doubling or period-tripling bifurcations as precursors to chaotic oscillations of ships in regular sea waves.

In another study Szemplinska-Stupnicka and Bajkowski (1986) have studied the Duffing oscillator of Ueda (3-2.25). They find subharmonic solutions using perturbation techniques and link the onset of chaos to the loss of stability of the subharmonics using classical stability analysis. They use analog computer experiments to check their results. They conclude that for the Duffing–Ueda attractor (3-2.35), the chaotic motion is a transition zone between the subharmonic and resonant harmonic solutions.

Although the author believes that the fundamental nature of chaotic motion is more closely related to such mathematical paradigms as horseshoe maps, fractals, and homoclinic orbits, the use of semiclassical methods of perturbation analysis may provide more practical analytic chaos criteria for certain classes of nonlinear systems.

5.4 LYAPUNOV EXPONENTS

Thus far we have discussed mainly predictive criteria for chaos. Here we describe a tool for *diagnosing* whether or not a system is chaotic. Chaos in deterministic systems implies a sensitive dependence on initial conditions. This means that if two trajectories start close to one another in phase space, they will move exponentially away from each other for small times on the average. Thus, if d_0 is a measure of the initial distance between the two starting points, at a small but later time the distance is

$$d(t) = d_0 2^{\lambda t} \qquad (5\text{-}4.1)$$

If the system is described by difference equations or a map, we have

$$d_n = d_0 2^{\Lambda n} \qquad (5\text{-}4.2)$$

[The choice of base 2 in Eqs. (5-4.1) and (5-4.2) is convenient but arbitrary.] The symbols Λ and λ are called *Lyapunov exponents*.[1]

---

[1] The terminology for Lyapunov exponents and numbers has not been standardized. Lichtenberg and Lieberman (1983) use the symbol σ for the exponent, Schuster (1985) uses λ, as does Wolf et al. (1985). But Guckenheimer and Holmes (1983) use μ for the exponent and Farmer et al. (1983) use λ to denote the Lyapunov *number*. Lyapunov was a Russian mathematician (1857–1918) who introduced this idea around the turn of the century.

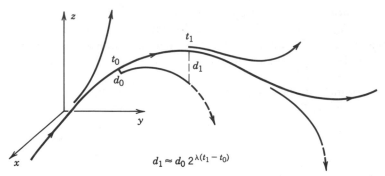

$$d_1 \approx d_0\, 2^{\lambda(t_1 - t_0)}$$

Figure 5-26 Sketch of the change in distance between two nearby orbits used to define the largest Lyapunov exponent.

An excellent review of Lyapunov exponents and their use in experiments to diagnose chaotic motion is given by Wolf et al. (1985). This review also contains two useful computer programs for calculating Lyapunov exponents.

The divergence of chaotic orbits can only be locally exponential, since if the system is bounded, as most physical experiments are, $d(t)$ cannot go to infinity. Thus, to define a measure of this divergence of orbits, we must average the exponential growth at many points along a trajectory, as shown in Figure 5-26. One begins with a reference trajectory [called a *fuduciary* by Wolf et al. (1985)] and a point on a nearby trajectory and measures $d(t)/d_0$. When $d(t)$ becomes too large (i.e., the growth departs from exponential behavior), one looks for a new "nearby" trajectory and defines a new $d_0(t)$. One can define the first Lyapunov exponent by the expression

$$\lambda = \frac{1}{t_N - t_0} \sum_{k=1}^{N} \log_2 \frac{d(t_\kappa)}{d_0(t_{\kappa-1})} \qquad (5\text{-}4.3)$$

Then the criterion for chaos becomes

$$\lambda > 0 \quad \text{chaotic}$$
$$\lambda \leq 0 \quad \text{regular motion} \qquad (5\text{-}4.4)$$

The reader by now has surmised that this operation can only be done with the aid of a computer whether the data are from a numerical simulation or from a physical experiment.

Only in a few pedagogical examples can one calculate λ explicitly. To examine one such case, consider the extension of the concept of Lyapunov exponents to a one-dimensional map

$$x_{n+1} = f(x_n) \tag{5-4.5}$$

In regions where $f(x)$ is smooth and differentiable, the stretch between neighboring orbits is measured by $|df/dx|$. To see this, suppose we consider two initial conditions x_0 and $x_0 + \epsilon$. Then in Eq. (5-4.2)

$$d_0 = \epsilon$$

$$d_1 = f(x_0 + \epsilon) - f(x_0) \simeq \frac{df}{dx}\bigg|_{x_0} \epsilon \tag{5-4.6}$$

Following Eq. (5-4.3), we define the Lyapunov or characteristic exponent as

$$\Lambda = \lim_{N \to \infty} \frac{1}{N} \sum_{k=0}^{N} \log_2 \left| \frac{df(x_n)}{dx} \right| \tag{5-4.7}$$

An illustrative example is the Bernoulli map

$$x_{n+1} = 2x_n \quad (\text{mod } 1) \tag{5-4.8}$$

as shown in Figure 5-27. Here (mod 1) means

$$x(\text{mod } 1) = x - \text{Integer}(x)$$

This map is multivalued and is known to be chaotic. Except for the switching value at $x = \frac{1}{2}$, $|f'| = 2$. Applying the definition (5-4.7), we find $\Lambda = 1$. Thus, on the average, the distance between nearby points grows as

$$d_n = d_0 2^n$$

The units of Λ are one bit per iteration. One interpretation of Λ is that one bit of information about the initial state is lost every time the map is iterated. To see this, write x_n in binary notation. For example, $x_n = (\frac{1}{2} + \frac{1}{4} + \frac{1}{16} + \frac{1}{128}) \equiv 0.1101001$ and x (mod 1) means 1.101001 (mod 1) = 0.101001. Thus, the map $2x_n$ (mod 1) moves the decimal point to the right and drops the integer value. So if we start out with m significant decimal places of information, we lose one for each iteration; that is, we lose one bit of information.

After m iterations we have lost knowledge of the initial state of the system.

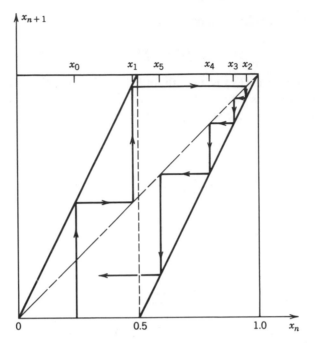

Figure 5-27   Chaotic orbit for the Bernoulli map, $X_{n+1} = 2X_n \pmod 1$.

Earlier in this chapter, we learned that the solution for the logistic or quadratic map becomes chaotic when the control parameter $a > 3.57$:

$$x_{n+1} = ax_n(1 - x_n) \tag{5-4.9}$$

This can be verified by calculating the Lyapunov exponent as a function of a as shown in Figure 5-28. Beyond $a = 3.57$, the exponent becomes nonpositive in the periodic windows $3.57 < a < 4$. When $a = 4$, it has been shown that $\lambda = \ln 2$ (e.g., see Schuster, 1984).

Another example of a map for which one can calculate the Lyapunov exponent is the *Tent map*

$$x_{n+1} = 2rx_n \qquad\qquad x_n < \tfrac{1}{2}$$

$$x_{n+1} = 2r - 2rx_n \qquad x \geq \tfrac{1}{2}$$

As in the Bernoulli map (5-4.8), $|f'(x)| = 2r$ is a constant and the Lyapunov exponent is found to be (Lichtenberg and Lieberman, 1983,

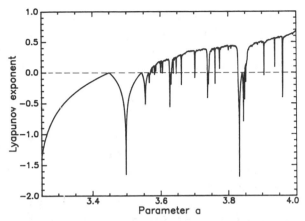

Figure 5-28 Lyapunov exponent versus control parameter a for the logistic equation (5-4.9).

pp. 416–417)

$$\lambda = \log 2r$$

When $2r > 1$, $\lambda > 0$ and the motion is chaotic, but when $2r < 1$, $\lambda < 0$ and the orbits are regular; in fact, all points in $0 < x < 1$ are attracted to $x = 0$ (Schuster, 1984, p. 22).

Numerical Calculation of the Largest Lyapunov Exponent

For every dynamical process, be it a continuous time history or discrete time evolution, there is a spectrum of Lyapunov or characteristic numbers that tells how lengths, areas, and volumes change in phase space. The idea of a spectrum of such numbers is discussed in the following section. However, in as far as a criterion for chaos is concerned, one need only calculate the largest exponent, which tells whether nearby trajectories diverge ($\lambda > 0$) or converge ($\lambda < 0$) on the average. As yet there is no analog instrument that will measure the Lyapunov exponent, although if this measure of chaotic motion continues to prove useful, some clever person will probably invent one. At the present time, however, calculations of Lyapunov exponents must be done by digital computer, preferably a midsized laboratory computer such as a Digital Equipment Corporation MicroVax or similar sized unit. A few papers have reported results using fast PC desk top computers.

There are two general methods, one for data generated by a known set of differential or difference equations (flows and maps) and the second to be used for experimental time series data. The Wolf et al. (1985) paper discusses both methods, but our experience to date reveals that more research on finding a reliable algorithm for experimental data is needed. We will review briefly techniques for a set of differential equations of the form

$$\dot{x} = f(x; c) \qquad (5\text{-}4.10)$$

where x is a set of n state variables and c is a set of n parameters. More complete discussion of these techniques may be found in Shimada and Nagashima (1979), the works of Benettin et al. (see the 1980 reference for a complete list), and Ueda (1979).

The main idea in calculating using (5-4.3) is to be able to determine the length ratio $d(t_k)/d(t_{k-1})$. One method is to numerically integrate the above set of equations to obtain a reference solution $x^*(t; x_0)$, where x_0 is the initial condition. Then at each time step t_k integrate the equation again, using as an initial condition some nearby point $x^*(t_k) + \eta$. However, a more direct method is to use the equation to find the variation of trajectories in the neighborhood of the reference trajectory $x^*(t)$. That is, at each time step t_k we solve the variational equations

$$\dot{\eta} = A \cdot \eta \qquad (5\text{-}4.11)$$

where A is the matrix of partial derivatives $\nabla f(x^*(t_k))$. We note that, in general, the elements of A depend on time. However, if A were constant, the solution of $\eta(t)$ between $t_k < t < t_{k+1}$ would depend on the initial condition. If this initial condition is chosen at random, then it is likely to have a component that lies in the direction of the largest positive eigenvalue of A. It is the change in length in this direction that the largest Lyapunov exponent measures.

Thus, the numerical scheme goes as follows. Integrate (5-4.10) to find $x^*(t)$. Allow a certain time to pass before calculating $d(t)$ in order to get rid of transients. After all, we are assuming we are on a stable attracter. After the transients are judged to be small, begin to integrate (5-4.11) to find $\eta(t)$. One can choose $|\eta(0)| = 1$, but choose the initial direction to be arbitrary. Then numerically integrate $\dot{\eta} = A(x^*(t)) \cdot \eta$, taking into account the change in A through $x^*(t)$. [In practice one can integrate both (5-4.10) and (5-4.11) simultaneously.] After a given time interval $t_{k+1} - t_k = \tau$, take

$$\frac{d(t_{k+1})}{d(t_k)} = \frac{|\eta(\tau; t_k)|}{|\eta(0; t_k)|} \qquad (5\text{-}4.12)$$

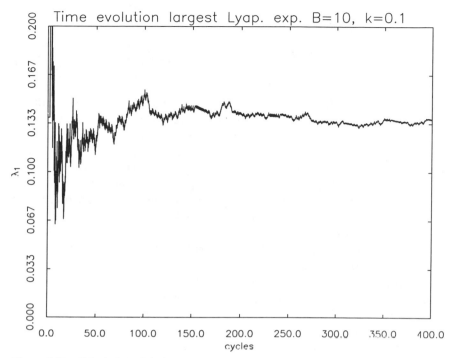

Figure 5-29 Calculation of the largest Lyapunov exponent for chaotic motion of the two-well potential attractor (5-3.26) as a function of the total time record.

To start the next time step in (5-4.3) use the direction of $\eta(\tau; t_k)$ for the new initial condition, that is,

$$\eta(0; t_{k+1}) = \frac{\eta(\tau; t_k)}{|\eta(\tau; t_k)|} \qquad (5\text{-}4.13)$$

where we have normalized the initial distance to unity.

An example of this calculation is shown in Figure 5-29 where we have numerically integrated the Duffing equation (1-2.4) in the chaotic state as a function of the elapsed time. The equations used were

$$\dot{x} = y$$

$$\dot{y} = -ky - x^3 + B \cos z \qquad (5\text{-}4.14)$$

$$\dot{z} = 1$$

The resulting matrix becomes

$$\mathbf{A} = \begin{bmatrix} \begin{bmatrix} 0 & 1 \\ -3x^2 & -k \end{bmatrix} & 0 \\ 0 & 0 & -B\sin z \end{bmatrix} \qquad (5\text{-}4.15)$$

Since this really is a periodically driven oscillator, changes of lengths in the phase space direction $z = t$ are zero, as manifested by the row of zeroes in the matrix $\mathbf{A}$. Thus, to find the largest Lyapunov exponent in this problem one can work in the *projection* of the phase space (x, y, z) onto the phase plane (x, y), using the inner bracketed matrix in (5-4.15).

For the data in Figure 5-29, the time step for numerical integration was $\Delta t = 0.01$ and the number of time steps to integrate $\eta(t)$ was chosen to be 10, or $\tau = 0.1$. The inner matrix in $\mathbf{A}$, (5-4.15), was updated at every Runge–Kutta time step Δt.

It is clear from Figure 5-29 that λ is a statistical property of the motion, that is, that one must average the changes in lengths over a long time in order to get a reliable value. Also, one has to be careful in choosing the Runge–Kutta step size Δt as well as the Lyapunov exponent step size τ.

A comparison of Lyapunov exponents for different parameters in the Duffing equation is shown in Table 5-1. This algorithm for calculating Lyapunov exponents has proved very useful in constructing empirical chaos criteria or chaos diagrams. If one has access to a really fast computer such as the so-called supercomputers, then one can calculate λ as a function of the parameters in the problem [c in (5-4.10)]. For example, one can choose $\mathbf{c} = (k, B)$ in the Duffing problem and find λ for 100×100 values of k and B. If $\lambda > 0$, then one prints out a symbol; otherwise, if $\lambda \sim 0$ or $\lambda < 0$, one leaves a blank. Such numerically determined chaos diagrams are useful to search for possible regions of parameter space where chaotic

TABLE 5-1   Comparison of Calculated Lyapunov Exponent for Duffing's Equation
$\ddot{x} + k\dot{x} + x^3 = B\cos t$

		(this book)[a]	(Ueda, 1979)	
k	B	λ_1	λ_1	λ_2
0.1	9.9	0.012	0.065	−0.166
0.1	10	0.094	0.102	−0.202
0.1	11	0.114	0.114	−0.214
0.1	12	0.143	0.149	−0.249
0.1	13	0.167	0.182	−0.282
0.1	13.3	0.174	0.183	−0.284

[a] Runge–Kutta integration time step, $\Delta t = 0.01$; Lyapunov restart time = $10\Delta t$; total time = 400 cycles = $800\ \pi$.

motion may exist (see Figure 5-3). Given the vagaries of numerical calculation however, one should not rely soley on this technique to certify a region as chaotic. Other tests such as spectral analysis, Poincaré maps, or fractal dimension should also be used to confirm suspected regions of chaotic motion.

Lyapunov Exponents and Distribution Functions. The calculation of the Lyapunov exponent (5-4.3) may be thought of as an average over time or iterates of the mapping (5-4.5). If one has a probability density function that tells the probability that certain trajectories will be in a certain region of phase space, then it is possible to replace this time average by a spatial average in phase space. This idea has been explored by several researchers (Everson, 1986; Hsu, 1986). The idea is illustrated for a two-dimensional map following Everson.

We recall that when the system is chaotic at least one Lyapunov exponent will be greater than zero. Start with the distance between two neighboring trajectories $\mathbf{x}_n$ and $\mathbf{y}_n$. This distance is given by $d_n = |\mathbf{x}_n - \mathbf{y}_n|$ and the Lyapunov exponent is given by

$$\Lambda = \lim_{N \to \infty} \frac{1}{N} \sum^{N} \log \frac{d_{n+1}}{d_n} \tag{5-4.16}$$

If an invariant probability distribution function $\rho(\mathbf{x})$ is assumed, then Λ can be calculated by

$$\Lambda = \iint \log \frac{d_{n+1}}{d_n} \rho(u, v)\, du\, dv \tag{5-4.17}$$

where a two-dimensional phase space is assumed with $\mathbf{x} = (u, v)$.

The invariant density function is assumed to satisfy the normalization condition

$$\iint \rho(u, v)\, du\, dv = 1,$$

where the integral is taken over all of phase space.

Everson applies this idea to a map related to the bouncing ball problem (3-2.9) and the standard map (5-3.32),

$$\theta_{n+1} = \theta_n + BV_n, \quad \mathrm{mod}\ 2\pi$$
$$V_{n+1} = \epsilon V_n + (1 + \epsilon)(1 + \sin \theta_{n+1}) \tag{5-4.18}$$

This is similar to the problem examined by Holmes (1982) where $0 < \epsilon < 1$ represents dissipation and BV_n represents the velocity of the ball as it leaves the platform at the nth bounce (see Figure 3-5a).

Everson uses two observations to apply (5-4.17) to (5-4.18) to calculate the largest Lyapunov exponent. First, he notes that from numerical experiments the invariant distribution function appears to be independent of the phase θ, so that in polar coordinates (V, θ)

$$\int_0^\infty \rho \, dV = \frac{1}{2\pi} \tag{5-4.19}$$

Second, he was able to obtain an approximate expression for the expression d_{n+1}/d_n, that is, for $B \gg 1$,

$$\frac{d_{n+1}}{d_n} \to |B(1 + \epsilon)\cos\theta| \tag{5-4.20}$$

which is independent of the velocity. Using (5-4.19) and (5-4.20) he was able to calculate

$$\Lambda = \log\frac{B(1 + \epsilon)}{2} \tag{5-4.21}$$

which agrees quite well with numerical calculations.

In another application of this technique, Hsu (1986) uses (5-4.17) but finds the probability density function numerically using a technique called cell mapping (e.g., see Hsu, 1981, 1987, as well as Kreuzer, 1985). Further study of the determination of invariant probability distribution functions in the future may allow more general application of this method of determining Lyapunov exponents.

Lyapunov Spectrum

Thus far we have talked only of the stretching of distance between orbits in a chaotic process. However, in three or more dimensions we know that regions of phase space may contract as well as stretch under a dynamic process. In particular, for dissipative systems, a small volume of initial conditions gets mapped into a smaller volume at a later time. This is illustrated in Figure 5-30 where a small sphere of initial conditions of radius δ is mapped at a later time into an ellipsoid with principal axes $(\mu_1^n\delta, \mu_2^n\delta, \mu_3^n\delta)$. Thus, to every dynamical system there is a spectrum of Lyapunov exponents or numbers $\{\lambda_i\}$, $\lambda_i = \log\mu_i$.

Computationally, this spectrum can be calculated from a time history of a motion in phase space by finding out how lengths, areas, volumes, and hypervolumes change under a dynamic process. Wolf et al. (1985) use this idea to develop a computation algorithm to calculate the $\{\lambda_i\}$. If the λ_i are ordered such that $\lambda_1 > \lambda_2 > \cdots > \lambda_n$, then they show that lengths vary as $d(t) \approx d_0 2^{\lambda_1 t}$, areas (formed from one point on the reference trajectory

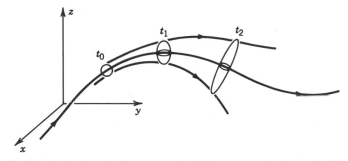

Figure 5-30   Sketch showing the divergence of orbits from a small sphere of initial conditions for a chaotic motion.

and two nearby points) vary as $A(t) \approx A_0 2^{(\lambda_1 + \lambda_2)t}$, and small volumes vary as $V(t) \approx V_0 2^{(\lambda_1 + \lambda_2 + \lambda_3)t}$ and so on.

Farmer et al. (1983) provide an analytic definition for the complete Lyapunov spectrum along with one example for which one can calculate the $\{\lambda_i\}$ exactly. In the remainder of this chapter we give a sketch of the calculation of Lyapunov exponents for a two-dimensional map. Many of the details are omitted and the interested reader is referred to the original Farmer et al. paper. To begin, we consider a general N-dimensional map

$$\mathbf{x}_{n+1} = \mathbf{F}(\mathbf{x}_n) \tag{5-4.22}$$

where $\mathbf{x}_n$ is a vector in an N-dimensional phase space. Then the change in shape of some small hypersphere will depend on the derivatives of the functions $\mathbf{F}(\mathbf{x}_n)$ with respect to the different components of $\mathbf{x}_n$. The relevant matrix is called a Jacobian matrix. For example, if

$$\mathbf{F} = (f(x, y, z), g(x, y, z), h(x, y, z))$$

then

$$J = \begin{bmatrix} \dfrac{\partial f}{\partial x} & \dfrac{\partial f}{\partial y} & \dfrac{\partial f}{\partial z} \\[2mm] \dfrac{\partial g}{\partial x} & \dfrac{\partial g}{\partial y} & \dfrac{\partial g}{\partial z} \\[2mm] \dfrac{\partial h}{\partial x} & \dfrac{\partial h}{\partial y} & \dfrac{\partial h}{\partial z} \end{bmatrix} = [\nabla \mathbf{F}] \tag{5-4.23}$$

After n iterations of the map, the local shape of the initial hypersphere depends on

$$[J_n] = [\nabla \mathbf{F}(\mathbf{x}_n)][\nabla \mathbf{F}(\mathbf{x}_{n-1})] \cdots [\nabla \mathbf{F}(\mathbf{x}_1)] \tag{5-4.24}$$

In general, one can find the eigenvalues of J_n which one orders according to

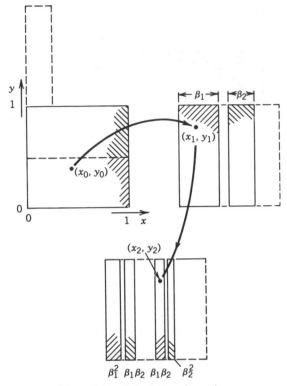

Figure 5-31   Baker's transformation.

$j_1(n) \geq j_2(n) \geq \cdots \geq j_N(n)$, where the $j_k(n)$ are the absolute values of the eigenvalues. The Lyapunov exponents are then defined by

$$\lambda_i = \lim_{n \to \infty} \frac{1}{n} \log_2 j_i(n) \qquad (5\text{-}4.25)$$

Farmer et al. illustrate the use of this definition for a two-dimensional map called a *baker's transformation* (Figure 5-31, named for its analogy to rolling and cutting pie dough. It is similar to the horseshoe map described in Chapter 1. The equations for this map are

$$x_{n+1} = \begin{cases} \lambda_a x_n & y < \frac{1}{2} \\ \frac{1}{2} + \lambda_b x_n & y > \frac{1}{2} \end{cases}$$

$$\qquad (5\text{-}4.26)$$

$$y_{n+1} = \begin{cases} 2y_n & y < \frac{1}{2} \\ 2\left(y - \frac{1}{2}\right) & y > \frac{1}{2} \end{cases}$$

This map is a generalization of the Bernoulli map in the previous section (5-4.8). In this case, the Jacobian matrix becomes

$$J = \begin{bmatrix} S_1 & 0 \\ 0 & 2 \end{bmatrix} \tag{5-4.27}$$

where $S_1 = \lambda_a$ for $y < \frac{1}{2}$ and $S_1 = \lambda_b$ for $y > \frac{1}{2}$.

For iterations of the map, the magnitudes of the eigenvalues become

$$j_1(n) = 2^n \qquad j_2(n) = \lambda_a^k \lambda_b^l, \qquad k + l = n$$

where one assumes that there are k iterations in the left half-plane and l iteration in the right half-plane. Applying the definition (5-4.25),

$$\lambda_1 = \lim_{n \to \infty} \frac{1}{n} \log_2 2^n$$

$$\lambda_2 = \lim_{n \to \infty} \left\{ \frac{k}{n} \log_2 \lambda_a + \frac{l}{n} \log_2 \lambda_b \right\}$$

TABLE 5-2   Lyapunov Exponents for Dynamical Models

System	Parameter Values	Lyapunov Spectrum (bits/s)	Lyapunov Dimensional (see Chapter 6)
Hénon		$\lambda_1 = 0.603$	
$X_{n+1} = 1 - aX_n^2 + Y_n$	$\begin{cases} a = 1.4 \\ b = 0.3 \end{cases}$	$\lambda_2 = -2.34$	1.26
$Y_{n+1} = bX_n$		(bits/iteration)	
Rossler chaos			
$\dot{X} = -(Y + Z)$	$a = 0.15$	$\lambda_1 = 0.13$	
$\dot{Y} = X + aY$	$b = 0.20$	$\lambda_2 = 0.00$	2.01
$\dot{Z} = b + Z(X - c)$	$c = 10.0$	$\lambda_3 = -14.1$	
Lorenz			
$\dot{X} = \sigma(Y - X)$	$\sigma = 16.0$	$\lambda_1 = 2.16$	
$\dot{Y} = X(R - Z) - Y$	$R = 45.92$	$\lambda_2 = 0.00$	2.07
$\dot{Z} = XY - bZ$	$b = 4.0$	$\lambda_3 = -32.4$	
Rossler hyperchaos			
$\dot{X} = -(Y + Z)$	$a = 0.25$	$\lambda_1 = 0.16$	
$\dot{Y} = X + aY + W$	$b = 3.0$	$\lambda_2 = 0.03$	3.005
$\dot{Z} = b + XZ$	$c = 0.05$	$\lambda_3 = 0.00$	
$\dot{W} = cW - dZ$	$d = 0.5$	$\lambda_4 = -39.0$	

Source: Wolf et al. (1985).

Here we invoke an assumption that after many iterations an orbit spends as much time in the left half-plane as in the right half-plane, or

$$\frac{k}{n} = \frac{1}{2} \qquad \frac{l}{n} = \frac{1}{2}$$

so that

$$\lambda_1 = 1, \qquad \lambda_2 = \tfrac{1}{2}\log_2\lambda_a\lambda_b < 0 \qquad\qquad (5\text{-}4.28)$$

Knowing these two Lyapunov exponents, one can then calculate a fractal dimension for this map. The relation between Lyapunov exponents and fractal dimensions has been examined by Farmer et al. (1983) and is discussed briefly in Chapter 6.

The spectra of Lyapunov exponents for several dynamics flows and maps are shown in Table 5-2 taken from the paper of Wolf et al. (1985).

As a final comentary on chaos criteria, we recently came across the following piece by the Italian bard F. Luna:

A Double Western

There was a man named Feigenbaum,
Who rode a bi-cycle into town.
The rutted road was so periodic,
His bike changed to a 4-cycle while on it.
The local sheriff, a man named Smale,[2]
Threw poor Feigenbaum into jail.
Judge Holmes then fined him for his trouble,
And warned him not to period double.
Said Holmes to Mitchell "We're gonna force you,
To ride your bi-cycle with proper Horseshoes'.
What is this town said the physicist so famous,
Said Smale its called by the name of Chaos.

F. LUNA

---

[2]Stephen Smale is a mathematician who in 1962–1963 continued the work of Poincaré and Birkoff and proved theorems relating the horseshoe map to homoclinic orbits and chaos (see Guckenheimer and Holmes, 1983, for a discussion of this work).

6

Fractal Concepts in Nonlinear Dynamics

Do you see O my brothers and sisters? It is not chaos or death
—it is form, union, plan—it is eternal life—it is Happiness.
                    Walt Whitman, *Leaves of Grass*

6.1 INTRODUCTION

Both "chaotic" and "strange attractor" have been used to describe the nonperiodic, randomlike motions that are the focus of this book. Whereas chaotic is meant to convey a loss of information or loss of predictability, the term strange is meant to describe the unfamiliar geometric structure on which the motion moves in phase space. In Chapter 5, we described a quantitative measure of the chaotic or informational loss aspect of these motions using Lyapunov exponents. In this chapter, we describe a quantitative measure of the strangeness of the attractor. This measure is called the fractal dimension. To do this, we have to describe the concept of fractal as it pertains to our applications.

In addition to the application of fractal ideas to a description of the dynamic attractor itself, it has been discovered that other geometric objects in the study of chaos, such as the boundary between chaotic and periodic motions in initial condition or parameter space, may also have fractal properties. Thus, we also include a section on *fractal basin boundaries*.

At the beginning of this book, we noted that the revolution in nonlinear dynamics has been sparked by the introduction of new geometric, analytic, and topological ideas which have given experimentalists (including numerical analysts) new tools to analyze dynamical processes. This in some ways parallels the earlier Newtonian revolution which introduced the calculus into dynamics. (Of course, Newton contributed much more by proposing new physical laws along with new mathematics.) Thus, in some sense, we are entering the second phase of the Newtonian revolution in dynamics and new geometric concepts like fractals must be mastered if one is to use the results of the new dynamics in practical problems.

Perhaps the most singular characteristic of chaotic vibrations in dissipative systems is the Poincaré map. These pictures provide a cross section of the attractor on which the motion rides in phase space and when the motion is chaotic, a mazelike, multisheeted structure appears. We have learned that this threadlike collection of points seems to have further structure when examined on a finer scale. To characterize such Poincaré patterns, we have used the term *fractal*. In this chapter, we try to make the mathematical meaning of fractal more precise. However, this treatment is not rigorous. Instead, what follows is one engineer's attempt to understand fractal structures and how to apply them to chaotic dynamics.

In the following section, we begin with a few simple examples of fractal curves and sets; namely, *Koch curves* and *Cantor sets*. We also introduce a quantitative measure of fractal qualities: the fractal dimension. Then we illustrate these concepts in several applications in nonlinear and chaotic vibrations.

The author presumes that the reader has no prior knowledge of set theory or topology beyond engineering mathematics at the baccalaureate level.

Koch Curve

This example is chosen from the book by Mandelbrot (1977) and was originally described by von Koch in 1904. One begins with a geometric construction that starts with a straight line segment of length 1. After dividing the line into three segments, one replaces the middle segment by two lines of length 1/3 as shown in Figure 6-1. Thus, we are left with four sides, each of length 1/3, so that the total length of the new boundary is 4/3. To get a fractal curve, one repeats this process for each of the new four segments and so on. At each step, the length is increased by 4/3 so that the total length approaches infinity. After many steps, one can see that the curve looks fuzzy. In fact, in the limit one has a continuous curve that is nowhere differentiable. In some sense, this new curve is trying to cover an area as a young child scribbling with crayons. Thus, we have the apparent paradox of a continuous curve that has some properties of an area. It is not

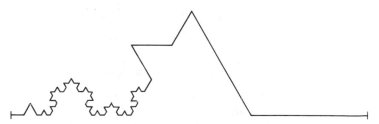

Figure 6-1   Partial construction of a fractal Koch curve.

surprising that one can define a dimension of this fractal curve which results in a value between 1 and 2.

Cantor Set

The Cantor set is attributed to George Cantor (1845–1918), who discovered it in 1883. It is a very important concept in modern nonlinear dynamics. If the Koch curve can be considered a process of adding finer and finer length structure to an initial line segment, then the Cantor set is the complement operation of removing smaller and smaller segments from a set of points initially on a line.

The construction begins as in the previous example with a line segment of length 1 which is subdivided into three sections as in Figure 6-2. However, instead of adding two more segments as in the Koch curve, one removes the middle segment of points so that the total number of segments is increased to two, and the total length is reduced to 2/3. This process is continued for the remaining line segments and so on. At each stage one throws away the middle segments of points creating twice as many line segments but reducing the total length by 2/3. In the limit, the total length

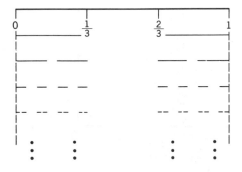

Figure 6-2 Top to bottom: sequential steps in the construction of a Cantor set.

approaches zero, although as we shall see below, the fractal dimension of this set of points is between zero and one.

The Devil's Staircase

The discontinuous fractal Cantor set can be used to generate a continuous fractal function by integrating an appropriate distribution function defined on the set. For example, suppose we imagine a distribution of mass on the interval $0 \le x \le 1$ with total mass equal to 1 in some units. Then if we redistribute the mass on the remaining Cantor intervals, at each step of the limiting process the mass density increases on the decreasing Cantor intervals such that the total mass is 1. At the nth step, the number of intervals is 2^n each of length $(1/3)^n$ so that the density is $(3/2)^n$. Integrating the mass density, we obtain the mass as a function of x:

$$M_n(x) = \int_0^x \rho_n(x)\, dx$$

where $\rho_n = (3/2)^n$ on the Cantor intervals and $\rho_n = 0$ otherwise. The limit of this process as $n \to \infty$ is a function called the devil's staircase, which has an infinite number of steps. One intermediate function $M_n(x)$ is shown in Figure 6-3.

In the limit, $M(x) = \text{limit}(n \to \infty)M_n(x)$. The expression $dM(x)/dx$ is an infinite set of delta functions.

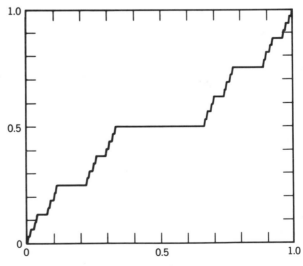

Figure 6-3   Devil's staircase function.

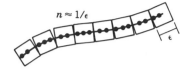

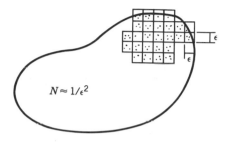

Figure 6-4 Covering procedure for linear and planar distributions of points.

Fractal Dimension

Thus far we have two examples of fractal sets but we do not have any test to determine if a set of points is fractal. To classify the Poincaré map of some nonlinear system, we need some quantitative measure of the fractal nature of the attractor.

There are many measures of the dimension of a set of points. We will describe a very intuitive or geometric definition called the *capacity*. Other definitions, which incorporate deeper mathematical subtleties, may be found in Mandelbrot[1] (1977) or Farmer et al. (1983), as well as in the next section. We begin with the measurement of the dimension of points along a line or distributed on some area.

First, consider a *uniform* distribution of N_0 points along some line or one-dimensional manifold in a three-dimensional space, as shown in Figure 6-4. We then ask how we can *cover* this set of points with small cubes with sides of length ϵ. (One can also use spheres of radius ϵ.) To be more specific, we calculate the minimum number of such cubes $N(\epsilon)$ to cover the set ($N(\epsilon) < N_0$). When N_0 is large, the number of cubes to cover a line will scale as

$$N(\epsilon) \approx \frac{1}{\epsilon}$$

Similarly, if we distribute points uniformly on some two-dimensional surface in three-dimensional space, we find that the minimum number of

[1] B. Mandlebrot is a mathematician with IBM Corp., Yorktown Heights, New York.

cubes to cover the set will scale in the following way:

$$N(\epsilon) \approx \frac{1}{\epsilon^2}$$

If the reader is convinced that this is intuitive, then it is natural to define the dimension by the following scaling law:

$$N(\epsilon) \approx \frac{1}{\epsilon^d} \qquad (6\text{-}1.1)$$

Taking the logarithm of both sides of Eq. (6-1.1) and adding a subscript to denote *capacity dimension*, we have

$$d_c = \lim_{\epsilon \to 0} \frac{\log N(\epsilon)}{\log(1/\epsilon)} \qquad (6\text{-}1.2)$$

Implicit in this definition is the requirement that the number of points in the set be large or $N_0 \to \infty$.

A set of points is said to be fractal if its dimension is noninteger—hence the term *fractal dimension*.

In the two examples of the Koch curve or Cantor set, the fractal dimension can be calculated exactly. For example, consider the nth iteration of the generation of the Koch curve where we let the size of the cubes be equal to the length of a straight line segment. At the nth step in the construction, the number of segments is

$$N_n = 4^n$$

while the size ϵ is given by

$$\epsilon_n = \left(\frac{1}{3}\right)^n$$

Replacing the limit $\epsilon \to 0$ with $n \to \infty$ in Eq. (6-1.2), one can easily see that for the *Koch curve*

$$d_c = \frac{\log 4}{\log 3} = 1.26185 \cdots \qquad (6\text{-}1.3)$$

Similarly, one can show that for the *Cantor set*

$$d_c = \frac{\log 2}{\log 3} \simeq 0.63092 \cdots \qquad (6\text{-}1.4)$$

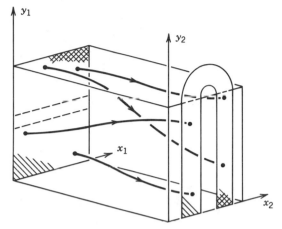

Figure 6-5   Horseshoe map.

One way to interpret the fractal dimension of the Koch curve is that the distribution of points covers more than a line but less than an area.

Two other examples of sets of which one can calculate the fractal dimension are the horseshoe map and the baker's transformation.

The *horseshoe map* has been discussed in Chapters 1 and 5 and is shown graphically in Figure 6-5. It is perhaps the simplest example of an iterative dynamical process in the plane that leads to a loss of information and fractal properties.

The calculation of the capacity fractal dimension for the horseshoe map is similar to that for the Cantor set except that the vertical direction leads to a contribution of "one" to the dimension. Using the definition (6-1.2), one can show that

$$d_c = \frac{\log 2}{\log|\epsilon|} + 1 \qquad (6\text{-}1.5)$$

where ϵ is the contraction parameter and $0 < \epsilon < 1/2$. [See also Bergé et al. (1985) for a discussion of this example.]

Another example for which one can calculate the fractal properties is the *baker's transformation* two-dimensional map. This example may be found in Farmer et al. (1983) and is similar to the horseshoe map (Figure 6-5). Its name derives from the idea of a baker rolling, stretching, and cutting pastry dough as shown in Figure 6-6. In this example, one can write out the specific difference equation relating a piece of dough at position (x_n, y_n) to

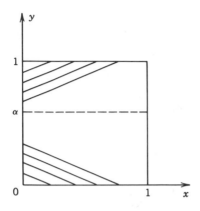

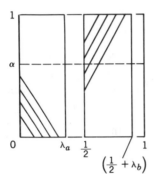

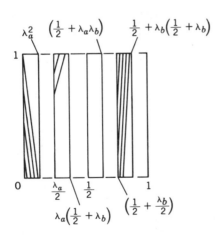

Figure 6-6 Baker's transformation or map.

it's new position in one iteration:

$$x_{n+1} = \begin{cases} \lambda_a x_n & \text{if } y_n < \alpha \\ \frac{1}{2} + \lambda_b x_n & \text{if } y_n > \alpha \end{cases}$$

$$y_{n+1} = \begin{cases} y_n/\alpha & \text{if } y_n < \alpha \\ \dfrac{1}{1-\alpha}(y_n - \alpha) & \text{if } y_n > \alpha \end{cases} \qquad (6\text{-}1.6)$$

where $0 \le x_n \le 1$ and $0 \le y_n \le 1$.

The article by Farmer et al. (1983) is very readable so we do not present the details but only quote the results. The problem is used by Farmer et al. to show the difference between different definitions of fractal dimension. They define the following function:

$$H(\alpha) = \alpha \log\frac{1}{\alpha} + (1 - \alpha)\log\frac{1}{1-\alpha} \qquad (6\text{-}1.7)$$

Using the definition of capacity, they find that

$$d_c = 1 + \bar{d}_c \qquad (6\text{-}1.8)$$

where $\bar{d}_c$ satisfies a transcendental equation

$$1 = \lambda_a^{\bar{d}_c} + \lambda_b^{\bar{d}_c} \qquad (6\text{-}1.9)$$

When $\lambda_a = \lambda_b = \lambda$,

$$d_c = 1 + \frac{\log 2}{\log|\lambda|} \qquad (6\text{-}1.10)$$

which is independent of α and identical to that for the horseshoe map (6-1.5).

It is probably safe to say that artists have intuitively understood the nature of fractal sets, especially the impressionists in the way in which they have used dots of color to achieve different effects of filling Euclidean space. In a more recent example, an advertisement in a popular magazine featured a Japanese artist whose design for a kimono material shows these fractal properties quite clearly (Figure 1-26).

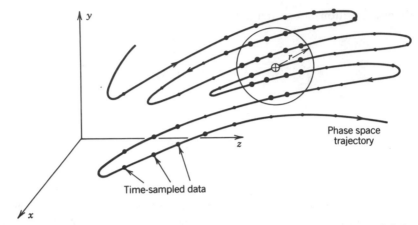

Figure 6-7 Long-time trajectory of motion in phase space showing the time-sampled data points and the counting sphere.

6.2 MEASURES OF FRACTAL DIMENSION

There are two criticisms of the use of capacity as a measure of fractal dimension of strange attractors—one theoretical and the other computational. First, capacity dimension is a geometric measure; that is, it does not account for the frequency with which the orbit might visit the covering cube or ball. Second, the process of counting a covering set of hypercubes in phase space is very time consuming computationally. In this section, we discuss three alternate definitions of fractal dimension which will address the shortcomings of the capacity. However, it should be pointed out that for many strange attractors these different dimensions give roughly the same value.

Pointwise Dimension

Let us consider a long time trajectory in phase space as shown in Figure 6-7. First, we time sample the motion so that we have a large number of points per orbit. Second, we place a sphere or cube of radius or length r at some point on the orbit and count the number of points within the sphere $N(r)$. The probability of finding a point in this sphere is then found by dividing by the total number of points in the orbit N_0; that is,

$$P(r) = \frac{N(r)}{N_0} \tag{6-2.1}$$

For a one-dimensional orbit, such as a closed periodic orbit $P(r)$ will be linear in r as $r \to 0$, $N_0 \to \infty$; $P(r) \approx br$. If the orbit were quasiperiodic, for example, it lay on a two-dimensional toroidal surface in a three-dimensional phase space, then the probability of finding a point on the orbit in a small cube or sphere of radius r would be $P(r) \approx br^2$. This leads one to define a dimension of an orbit at a point x_i (here x_i is a vector in phase space) by measuring the relative percentage of time that the orbit spends in the small sphere; that is,

$$d_P = \lim_{r \to 0} \frac{\log P(r; x_i)}{\log r} \qquad (6\text{-}2.2)$$

For some attractors, this definition will be independent of the point x_i. But for many, d_P will depend on x_i and an averaged pointwise dimension is best used. Also, for some sets of points such as a Cantor set, there will be gaps in the distribution of points so that $P(r)$ is not a smooth function of r as $r \to 0$, as can be seen in the devil's staircase in Figure 6-3.

To obtain an averaged pointwise dimension, one randomly chooses a set of points $M < N_0$ and calculates $d_P(x_i)$ at each point. The averaged pointwise dimension is then given by

$$\hat{d}_P = \frac{1}{M} \sum_{i=1}^{M} d_P(x_i) \qquad (6\text{-}2.3)$$

As an alternative, one can average the probabilities $P(r; x_i)$. Choose a random *subset* of M points distributed around the attractor, where $M \ll N_0$. We then conjecture that

$$\lim_{r \to 0} \frac{1}{M} \sum_{i=1}^{M} P(r; x_i) = ar^{d_P}$$

or

$$d_P = \lim_{r \to 0} \frac{\log(1/M)\Sigma P(r)}{\log r}$$

In practice, if $N_0 \approx 10^3\text{--}10^4$ points, then $M \approx 10^2\text{--}10^3$.

Another method for calculating the fractal dimension involves averaging over the radii of balls or sizes of cubes in phase space which contain the same number of points, say N. Choosing different reference points $\{x_i\}$ one finds $r_i(N)$ and takes the average over n reference points

$$\bar{r}(N) = \frac{1}{n} \sum_{i=1}^{n} r_i(N)$$

Then one assumes that the fractal scaling goes as

$$N = c\bar{r}^d$$

This method has been studied by Termonia and Alexandrowicz (1983).

Correlation Dimension

This measure of fractal dimension has been used successfully by experimentalists (e.g., see Malraison et al., 1983; Swinney, 1985; Ciliberto and Gollub, 1985; and Moon and Li, 1985a) and in some ways is related to the pointwise dimension. An extensive study of this definition of dimension has been given by Grassberger and Proccacia (1983).

As in the definition of pointwise dimension, one discretizes the orbit to a set of N points $\{x_i\}$ in the phase space. (One can also create a pseudo-phase-space; see Chapter 4 and the next section.) One then calculates the distances between pairs of points, say $s_{ij} = |x_i - x_j|$ using either the conventional Euclidean measure of distance (square root of the sum of the squares of components) or some equivalent measure such as using the sum of absolute values of vector components. A correlation function is then defined as

$$C(r) = \lim_{N \to \infty} \frac{1}{N^2} \left(\begin{array}{c} \text{number of pairs } (i, j) \\ \text{with distance } s_{ij} < r \end{array} \right) \tag{6-2.4}$$

For many attractors this function has been found to exhibit a power law dependence on r as $r \to 0$; that is,

$$\lim_{r \to 0} C(r) = ar^d$$

so that one may define a fractal or correlation dimension using the slope of the $\ln C$ versus $\ln r$ curve:

$$d_G = \lim_{r \to 0} \frac{\log C(r)}{\log r} \tag{6-2.5}$$

It has been shown that $C(r)$ may be calculated more effectively by constructing a sphere or cube at each point x_i in phase space and counting the number of points in each sphere; that is,

$$C(r) = \lim_{N \to \infty} \frac{1}{N^2} \sum_i^N \sum_{\substack{j \\ i \ne j}}^N H(r - |x_i - x_j|) \qquad (6\text{-}2.6)$$

where $H(s) = 1$ if $s > 0$ and $H(s) = 0$ if $s < 0$. This differs from the pointwise dimension in that here the sum is performed about *every* point.

Information Dimension

Many investigators have suggested another definition of fractal dimension that is similar to the capacity (6-1.2) but tries to account for the frequency with which the trajectory visits each covering cube. As in the definition of capacity, one covers the set of points, whose dimension one wishes to measure, with a set of N cubes of size ϵ. This set of points is again a uniform discretization of the continuous trajectory. (It is assumed that a long enough trajectory is chosen to effectively cover the attractor whose dimension one wants to measure. For example if the motion is quasiperiodic, the trajectory has to run long enough to "visit" all regions on the toroidal surface of the attractor.)

To calculate the information dimension, one counts the number of points N_i in each of the N cells and the probability of finding a point in that cell P_i, where

$$P_i \equiv \frac{N_i}{N_0}, \qquad \sum_i^N P_i = 1 \qquad (6\text{-}2.7)$$

where N_0 is the total number of points in the set. Note that $N_0 \ne N$. The *information entropy* is defined by the expression

$$I(\epsilon) = -\sum_i^N P_i \log P_i \qquad (6\text{-}2.8)$$

[When the log function is with respect to base 2, $I(\epsilon)$ has the units of *bits*.] For small ϵ, it is found that I behaves as,

$$I \approx d_I \log(1/\epsilon)$$

so that for small ϵ we may define a dimension

$$d_I = \lim_{\epsilon \to 0} \frac{I(\epsilon)}{\log(1/\epsilon)} = \lim \frac{\Sigma P_i \log P_i}{\log \epsilon}. \qquad (6\text{-}2.9)$$

To see that this definition is related to the capacity, we note that if the probabilities P_i were equal for all cells, that is,

$$P_i = \frac{N_i}{N_0} = \frac{1}{N} \qquad (6\text{-}2.10)$$

then

$$I = \sum P_i \log P_i = -N \frac{1}{N} \log \frac{1}{N} = \log N$$

so that $d_I = d_c$. In general, it can be shown that (see Farmer et al., 1983)

$$d_I \leq d_c \qquad (6\text{-}2.11)$$

Further discussion of the information dimension may be found in Farmer et al. (1983), Grassberger and Proccacia (1983), and Shaw (1984).

The information entropy is a measure of the *unpredictability* in a system. That is, for a uniform probability in each cell, $N_i = 1/N$, I is at a maximum. If all the points are located in one cell (maximum predictability), $I = 0$, as can be seen by the calculation

For $P_i = 1/N$, $\quad I = \log N$

For $P_1 = 1$, $P_i = 0$, $i \neq 1$, $\quad I = 1 \cdot \log 1 = 0$

Definition (6-2.8) and use of the symbol $I(\epsilon)$ are confusing in the literature. Shaw (1984) uses the symbol H to denote entropy and I to denote the negative entropy $(-H)$ or *information*. Thus, for Shaw, a more predictable system (i.e., sharper P_i distribution) has *higher* information.

Relation Between Dimension Definitions and Lyapunov Exponents

Thus far we have defined the following fractal dimensions:

d_c the capacity (6-1.2)
d_P pointwise dimension (6-2.2)
d_G correlation dimension (6-2.5)
d_I information dimension (6-2.9)

TABLE 6-1   Fractal Dimension of Selected Dynamical Systems

Name of Systems	Dimension	Type	Source of Data
Henon map (1-3.8)	1.26	Capacity	Grassberger and Proccacia (1983)
($a = 1.4$, $b = 0.3$)	1.21 ± 0.01	Correlation	
Logistic map (1-3.6)	0.538	Capacity	Grassberger and Proccacia (1983)
($\lambda = 3.5699456$)	0.500 ± 0.005	Correlation	
Lorenz equations	2.06 ± 0.01	Capacity	Grassberger and Proccacia (1983)
(1-3.9)	2.05 ± 0.01	Correlation	
Two-well potential	$2.14\,(\gamma = 0.15)$	Correlation	Moon and Li (1985a)
(Eqn. 6-3.7, $f = 0.16$,			
$\omega = 0.8333$)	$2.61\,(\gamma = 0.06)$		
Chua's circuit	2.82	Lyapunov	Matsumoto et al. (1985)

Grassberger and Proccacia (1983) have shown that the information dimension and the correlation dimension are lower bounds on the capacity definition; that is,

$$d_G \le d_I \le d_c \tag{6-2.12}$$

For many of the standard strange attractors, however, all three were very close (see Table 6-1).

These same authors have also pointed out a connection between capacity, correlation function, and information entropy. In statistical mechanics and information theory, one can define a set of information entropy measures called *order-q information* (see Grassberger and Proccacia, 1984):

$$I_q = \frac{1}{1 - q}\log \sum_i^N P_i^q \tag{6-2.13}$$

where the P_i are the probability of finding points in a set of N covering cubes. If the size of the covering cubes is ϵ, one can define *order-q dimensions*

$$d_q = \lim_{\epsilon \to 0} \frac{I_q(\epsilon)}{\log(1/\epsilon)} \tag{6-2.14}$$

When $q = 0, 1, 2$, one can relate the corresponding d_q to the capacity, information, and correlation dimensions, respectively.

For $q = 0$

$$I_0 = \log \sum_i^N P_i^0 = \log N$$

For $q = 1$ (use $q = 1 + \eta$, let $\eta \to 0$)

$$I_1 = \lim_{\eta \to 0} \frac{1}{\eta} \log \sum P_i P_i^{\eta} = -\sum P_i \log P_i$$

For $q = 2$

$$I_2 = -\log \sum_i^N P_i^2 = \lim_{N_0 \to 0} \log 2 N_0 C(\epsilon)$$

Thus, the capacity dimension takes no account of the distribution of points between covering cells, whereas the order one information entropy dimension measures the probability of finding a point in a cell. Finally, the correlation dimension accounts for the probability of finding two points in the same cell.

A further relation between fractal dimension, information entropy, and Lyapunov exponents was made by Kaplan and Yorke (1978). We recall from Chapter 5 that the Lyapunov exponents measure the rate at which trajectories *on* the attractor diverge from one another and trajectories *off* the attractor converge toward the attractor (e.g., see Figure 5-30). Thus, a small sphere of initial conditions centered at some point on the attractor in phase space is imagined to deform in time under the dynamical process into an ellipse. For example, for a chaotic two-dimensional map,

$$\mathbf{x}_{n+1} = f(\mathbf{x}_n) \tag{6-2.15}$$

a circle of initial conditions (with radius ϵ) deforms into an ellipse after M iterations of the map. The major and minor radii are given by $L_1^M \epsilon$ and $L_2^M \epsilon$. When L_1 and L_2 are averaged over the whole attractor, they are referred to as Lyapunov numbers and $\lambda_i = \log L_i$ are called the Lyapunov exponents.

Kaplan and Yorke (1978) (see also Farmer et al., 1983)[2] have suggested that one can calculate a dimension for a fractal attractor based on the Lyapunov exponents. For a two-dimensional map, this dimension becomes

$$d_L = 1 + \frac{\log L_1}{\log(1/L_2)} = 1 - \frac{\lambda_1}{\lambda_2} \tag{6-2.16}$$

For higher-dimensional maps in an N-dimensional phase space, the relation is more complicated. First, we order the Lyapunov numbers; that is,

$$L_1 > L_2 > \cdots > L_k > \cdots > L_N \tag{6-2.17}$$

---

[2] Note that Farmer et al. (1983) use λ to denote the Lyapunov *number*, not the Lyapunov exponent.

Then we find L_k such that the product

$$L_1 L_2 \cdots L_k \geq 1$$

The Lyapunov dimension is defined to be

$$d_L = k + \frac{\log(L_1 L_2 \cdots L_k)}{\log(1/L_{k+1})} \tag{6-2.18}$$

Kaplan and Yorke (1978) suggest that this is a lower bound on the capacity dimension; that is,

$$d_L \leq d_c \tag{6-2.19}$$

As an example, consider a three-dimensional set of points generated by a Poincaré map of a fourth-order set of first-order differential equations with dissipation. If the attractor is strange, we find

$$L_1 > 1, \qquad L_2 = 1, \qquad L_3 < 1$$

For example, one principal axis of the ellipsoid of initial conditions grows, one stays the same length, and one axis contracts. Also, since the system is dissipative, the volume of the ellipsoid must be less than that of the original sphere of initial conditions so that $L_1 L_2 L_3 < 1$. This leads us to use $k = 2$ in Eq. (6-2.18) and

$$d_L = 2 + \frac{\log L_1}{\log(1/L_3)} = 2 + \frac{\lambda_1}{|\lambda_3|} \tag{6-2.20}$$

The usefulness of this formula for experimental data is unclear at this time since it is not easy to obtain a measurement of the contraction Lyapunov number L_3 (e.g., see Wolf et al., 1985).

A comparison of the different definitions of fractal dimension for the baker's transformation (6-1.6) has been given by Farmer et al. (1983). This example is one of the few dynamical systems for which one can analytically calculate the properties of the chaotic dynamics.

Using definition (5-4.13), they show that the Lyapunov dimension (6-2.20) is equal to the information dimension (6-2.9) and is given by

$$d_I = d_L = 1 + \frac{H(\alpha)}{\alpha \log(1/\lambda_a) + \beta \log(1/\lambda_b)} \tag{6-2.21}$$

where $\beta = 1 - \alpha$. When $\lambda_a = \lambda_b$, one can show that

$$d_I = d_L = 1 + \frac{H(\alpha)}{\log(1/\lambda)} \tag{6-2.22}$$

Furthermore, if $\alpha = \frac{1}{2}$, $H(\alpha) = \log 2$ and

$$d_I = d_1 = d_c$$

In some ways, α and λ_a/λ_b represent inhomogeneity factors in the map. When $\alpha = \frac{1}{2}$ and $\lambda_a/\lambda_b = 1$, the map is like the horseshoe or Cantor map and all these definitions of dimension d_I, d_L, d_c become equal. The implications are that different definitions of fractal dimension are likely to yield different results when the dynamical process leads to a "nonuniform" Poincaré map.

How Useful Is Fractal Dimension for Vibration Problems?

The practical use of all the dimensions discussed above in measuring and characterizing chaotic vibrations has yet to be fully settled. In many cases, it is sufficient to establish that the dimension is not integer or that the attractor is indeed strange. However, for some strange attractors the fractal dimension is close to an integer [e.g., for the Lorenz attractor (1-3.9), $d \approx 2.06$], so that the fractal dimension may not by itself establish the chaotic nature of the motion. As suggested in Chapter 2, it is better not to rely on one measure of chaos in dynamical experiments, but to use two or more techniques such as Poincaré maps, Fourier spectra, Lyapunov exponents, or fractal dimension measurements before pronouncing a system chaotic or strange.

In the next section, we discuss the application of fractal dimension in characterizing strange attractors.

6.3   FRACTAL DIMENSION OF STRANGE ATTRACTORS

There are two principal applications of fractal mathematics to nonlinear dynamics: characterization of strange attractors and measurement of fractal boundaries in initial condition and parameter space. In this section, we discuss the use of the fractal dimension in both numerical and experimental measurements of motions associated with strange attractors.

As yet, there are no instruments, electronic or otherwise, that will produce an output proportional to the fractal dimension, although electro-optical methods may achieve this end in the future (see Section 6.5). To date, in both numerical and experimental measurements, the fractal dimension and Lyapunov exponents are found by discretizing the signals at uniform time intervals and processing the data with a computer. There are three basic methods:

(a)   Time discretization of phase space variables
(b)   Calculation of fractal dimension of Poincaré maps
(c)   Construction of pseudo-phase-space using single variable measure-
      ments (sometimes called the embedding space method)

In both the first and third methods, the variables are measured and
stored at uniform time intervals $\{x(t_0 + n\tau)\}$, where n is a set of integers.
The time interval τ is chosen to be a fraction of the principal forcing period
or characteristic orbit time. If the Poincaré map in (b) is based on a time
signal, τ is just the period of the time-based Poincaré map. However, if the
Poincaré map is based on other phase space variables, the data are collected
at variable times depending on the specific type of Poincaré map (see
Chapter 4).

There are three principal definitions of fractal dimension used today:
averaged pointwise dimension, correlation dimension, and Lyapunov dimension. In most of the current experience with actual calculation of fractal
dimension, between 2000 and 20,000 points are used, although several
recent papers claim to have reliable algorithms based on as little as 500
points (e.g., see Abraham et al., 1986). Direct algorithms for calculating
fractal dimension based on N_0 points generally take N_0^2 operations so that
superminicomputers or mainframe computers are often used. However,
clever use of basic machine operations can reduce the number of operations
to order $N_0 \ln N_0$ and significantly speed up calculation (e.g., see
Grassberger and Proccacia, 1983).

(a)   Discretization of Phase Space Variables

Suppose we know or suspect a chaotic system to have an attractor
in three-dimensional phase space based on the physical variables
$\{x(t), y(t), z(t)\}$. For example, in the case of the forced motion of a beam
or particle in a two-well potential (see Chapter 2) $x =$ position, $v = \dot{x}$ is the
velocity, and $z = \omega t$ is the phase of the periodic driving force. In this
method, time samples of $(x(t), y(t), z(t))$ are obtained at a rate that is
smaller than the driving force period. To each time interval there corresponds a point $x_n = (x(n\tau), y(n\tau), z(n\tau))$ in phase space.

To calculate an averaged pointwise dimension, one chooses a number of
random points x_n. About each point one calculates the distances from x_n to
the nearest points surrounding x_n. (Note that these points are not the

nearest in time, but in distance.) One does not need to use a Euclidean measure of distance. For example, the sum of absolute values of the components of $(\mathbf{x}_n - \mathbf{x}_m)$ could be used, that is,

$$s_{nm} = |x(n\tau) - x(m\tau)| + |y(n\tau) - y(m\tau)| + |z(n\tau) - z(m\tau)| \quad (6\text{-}3.1)$$

Then the number of points within a ball, cube, or other geometric shape of order ϵ is counted and a probability measure is found as a function of ϵ.

$$P_n(\epsilon) = \frac{1}{N_0} \sum_{m=1} H(\epsilon - s_{nm}) \quad (6\text{-}3.2)$$

where N_0 is the total number of sampled points and H is the Heaviside step function: $H(r) = 1$ if $r > 0$; $H(r) = 0$ if $r < 0$. The averaged pointwise dimension, following Eq. (6-2.3), is then

$$d_n = \lim_{\epsilon \to 0} \frac{\log P_n(\epsilon)}{\log \epsilon}$$

$$(6\text{-}3.3)$$

$$d \equiv \frac{1}{M} \sum_{n=1}^{M} d_n$$

where the limit defining d_n exists. For some attractors, the function P_n versus ϵ is not a power law but has steps or abrupt changes in slope. One can then calculate a modified average pointwise dimension by first averaging P_n. For example, let

$$\hat{C}(\epsilon) = \frac{1}{M} \sum_{n=1}^{M} P_n(\epsilon)$$

$$(6\text{-}3.4)$$

$$d = \lim_{\epsilon \to 0} \frac{\log \hat{C}(\epsilon)}{\log \epsilon}$$

This is similar to the correlation dimension discussed in the previous section.

The example of the two-well potential (5-2.2) is shown in Figure 6-8a, b using the correlation dimension. This dimension is computed from numerically generated data using the equation $\dot{x} = y$, $\dot{y} = -sy - \frac{1}{2}x(1 - x^2) + f \cos z$, $\dot{z} = \omega$ for values of δ, f, ω in the chaotic regime. Figure 6-8a shows the logarithm of the correlation function while Figure 6-8b shows the local slope versus the logarithm of the size of the test volume. The slope for the intermediate values of ϵ is around 2.5. This is consistent with the fact that the attractor lives in a three-dimensional space (x, y, z).

In practice, $N_0 \approx 3\text{--}10^3 \times 10^4$ points and $M \approx .20N_0$. One should experiment with the choice of M by starting with a small value and increasing it until d reaches some limit.

The choice of ϵ also requires some judgment. The upper limit of ϵ is much smaller than the maximum size of the attractor yet large enough to capture the large-scale structure in the vicinity of the point x_n. The smallest value of ϵ must be such that the associated sphere or cube contains at least one sample point. For example, in a three-dimensional phase space, if the mean global scale of the attractor is L, the average point density is

$$\rho \approx \frac{N_0 6}{\pi L^3}$$

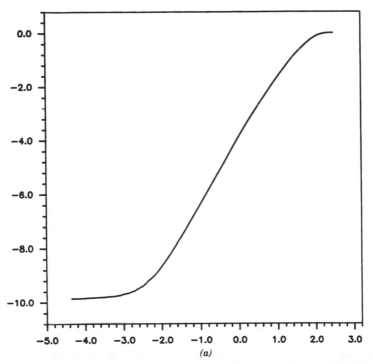

(a)

Figure 6-8 (a) log C versus log ϵ for chaotic motion in a two-well potential (3-3.6). Data obtained from numerical integration.

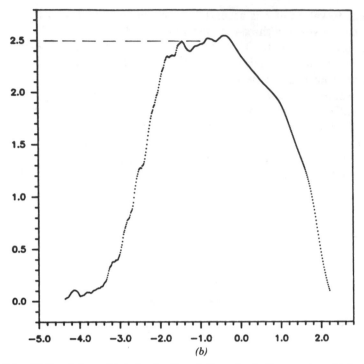

(b)

Figure 6-8 (*b*) Local slope of (*a*) showing fractal dimension in linear region of (*a*) of around 2.5.

so that the volume associated with ϵ should be greater than ρ^{-1} or

$$\epsilon > \frac{L}{2N_0^{1/3}} \qquad (6\text{-}3.5)$$

Another constraint on the minimum size of ϵ is the "real noise" or uncertainty in the measurements of the state variables (x, y, z). In an actual experiment, there is a sphere of uncertainty surrounding each measured point in phase space. When ϵ becomes smaller than the radius of this sphere, the theory of fractal dimension discussed above comes into question since for smaller ϵ one cannot expect a self-similar structure.

(b)   Fractal Dimension of Poincaré Maps

In systems driven by a periodic excitation, as in the Duffing–Ueda strange attractor (3-2.25) or the two-well potential strange attractor (3-3.6), time or the phase $\phi = \omega t$ becomes a natural phase space variable. In most cases, this time variable will lie in the attractor subspace and time can be considered as one of the contributions to the dimension of the attractor. In the case of a periodically forced, nonlinear, second-order oscillator, the Poincaré map based on periodic time samples produces a distribution of points in the plane. To calculate the fractal dimension of the complete attractor, it is sometimes convenient to calculate the fractal dimension of the Poincaré map $0 < D < 2$. If D is independent of the phase of the Poincaré map (remember $0 \le \omega t \le 2\pi$), the dimension of the complete attractor is just

$$d = 1 + D \qquad (6\text{-}3.6)$$

As an example, we present numerical and experimental data for the two-well potential or Duffing–Holmes strange attractor (Chapter 2):

$$\ddot{x} + \gamma \dot{x} - \tfrac{1}{2}x(1 - x^2) = f \cos \omega t \qquad (6\text{-}3.7)$$

In this example, we are interested in two questions:

1. Does the fractal dimension of the strange attractor vary with the phase of the Poincaré map?
2. How does the fractal dimension vary with the damping γ?

The fractal dimension was calculated for a set of experimental Poincaré maps and are listed in Table 6-2. This table shows an almost constant value around the attractor. Thus, the assumption $d = 1 + D$ in Eq. (6-3.6) appears to be a good one.

TABLE 6-2   Dimension of Experimental Poincaré Map versus Phase for Vibration of a Buckled Beam[a]

ϕ	$D(1,4)$[b]	$D(1,7)$[c]
0	1.741	1.628
45	1.751	1.627
90	1.742	1.638
135	1.748	1.637
180	1.730	1.637

[a]Nondimensional damping, $\gamma = 0.013$; forcing frequency, 8.5 Hz; natural frequency about buckled state, 9.3 Hz; from Moon and Li (1985a).
[b]Based on four smallest log r points in log C versus log r.
[c]Based on seven smallest log r points in log C versus log r.

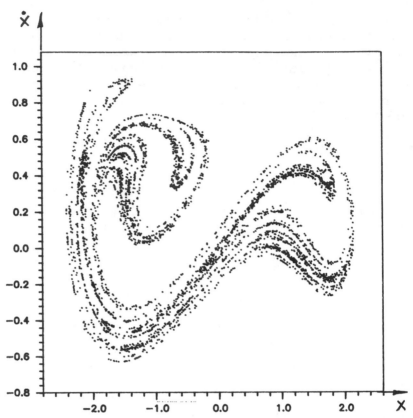

Figure 6-9   Fractal distribution of points from a Poincaré map for the two-well potential problem (3-2.10), using the same data as in Figure 6-8.

A numerically generated Poincaré map for the case of a particle in a two-well potential under periodic excitation is shown in Figure 6-9. The correlation function (Figure 6-10a $C(\epsilon)$ vs ϵ is shown plotted in a log-log scale and shows a linear dependence as assumed in the theory.

The data in Figure 6-10 was the same as that used in Figure 6-8. From Figure 6-10b, $D \simeq 1.5$ or $d = 2.5$, which agrees with that calculated directly from the attractor in the phase space $(x, \dot{x}, \omega t)$ as in Figure 6-8.

The effect of damping on the fractal dimension of the two-well potential strange attractor was determined from Runge–Kutta numerical simulation. This dependence is shown in Figure 6-11. The data show that low damping yields an attractor that fills phase space ($D = 2$, $d = 3$) as would a Hamiltonian (zero damping) system. As damping is increased, however, the Poincaré map looks one dimensional and the attractor has a dimension close to $d = 2$, as in the case of the Lorenz equations.

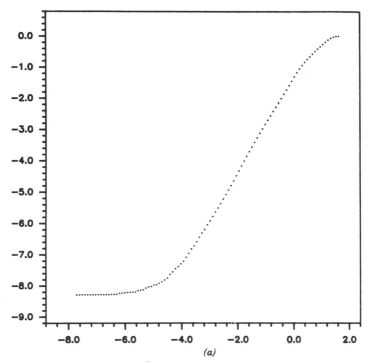

Figure 6-10   (*a*) log *C* versus log ϵ for the set of points in the Poincaré map in Figure 6-9.

The fractal dimension of a chaotic circuit (diode, inductor, and resistor in series driven with an oscillator) has been measured by Linsay (1985) using a Poincaré map. He measures the current at a sampling time equal to the period of the oscillator and constructs a three-dimensional pseudo-phase-space using $(I(t), I(t + \tau), I(t + 2\tau))$ (see next section). He obtains a fractal dimension of the Poincaré map of $D = 1.58$ and infers a dimension of the attractor of 2.58.

(c) Dimension Calculation from Single Time Series Measurement

The methods discussed above assume that (1) one knows the dimension of the phase space wherein the attractor lies and (2) one has the ability to measure all the state variables. However, in many experiments, the time history of only one state variable may be available or possible. Also, in continuous systems involving fluid or solid continua, the number of degrees of freedom or minimum number of significant modes contributing to the chaotic dynamics may not be known a priori. In fact, one of the important

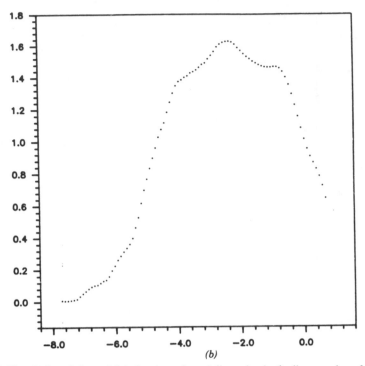

Figure 6-10   (*b*) Local slope of (*a*) showing a fractal dimension in the linear region of (*a*) of around 1.5.

applications of fractal mathematics is to allow one to determine the smallest number of first-order differential equations that may capture the qualitative features of the dynamics of continuous systems. This has already had some success in thermofluid problems such as Rayleigh–Benard convection (see Malraison et al., 1983).

In early theories of turbulence (e.g., Landau, 1944), it was thought that chaotic flow was the result of the interaction of a very large or infinite set of modes or degrees of freedom in the fluid. At the present time, it is believed that the chaos associated with the transition to turbulence can be modeled by a finite set of ordinary differential equations.

Thus, suppose that the number of first-order equations required to simulate the dynamics of a dissipative system is N. Then the fractal dimension of the attractor would be $d < N$. Then if we were to determine d by some means, we would then determine the minimum N.

Not knowing N, we cannot know how many physical variables $(x(t), y(t), z(t), \dots)$ to measure. Instead, we construct a pseudo-phase-space, or embedding space, using time-delayed measurements of one physi-

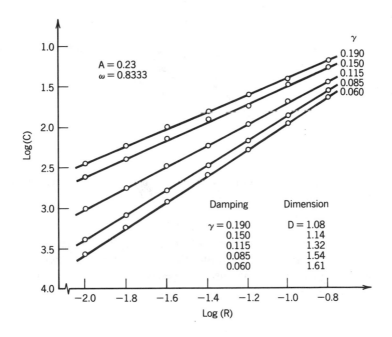

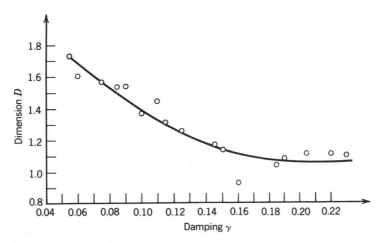

Figure 6-11 Dependence of fractal dimension on the damping for the two-well potential oscillator (3-2.10).

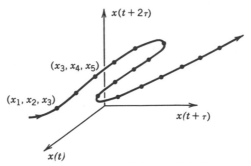

Figure 6-12   Sketch of an orbit in a three-dimensional pseudo-phase-space constructed from a single time series measurement.

cal variable, say $(x(t), x(t + \tau), x(t + 2\tau), \ldots)$ (see Chapter 4 and also Packard et al., 1980). For example, three-dimensional pseudo-phase-space vectors are calculated using three successive components of the digitized $x(t)$ (Figure 6-12), that is,

$$\mathbf{x}_n = \big\{ x(t_0 + n\tau), x(t_0 + (n + 1)\tau), x(t_0 + (n + 2)\tau) \big\} \quad (6\text{-}3.8)$$

With these position vectors, one can use the correlation function (6-2.6) or averaged probability function (6-2.3) to calculate a fractal dimension.

To determine the minimum N, one constructs higher-dimensional pseudo-phase-spaces based on the time-sampled $x(t)$ measurements until the value of the fractal dimension reaches an asymptote, say, $d = M + \mu$, where $\mu < 1$. Then the minimum phase space dimension for this chaotic attractor is $N = M + 1$.

In reconstructing a dynamical attractor from the time history measurements of a single variable, the question arises of how many dimensions are required in the embedding space in order to capture all the topological features of the original attractor. A mathematician named Takens has proved several theorems about this question. If the original phase space attractor lives in an N = dimensional space, then in general one must reconstruct an embedding space (our pseudo-phase-space) of $2N + 1$ dimensions.

To illustrate these ideas we have applied the embedding space method to find the dimension of the two-welled potential (or buckled beam) attractor (5-2.2). Earlier we saw that this attractor lives in a three-dimensional phase space $(x, \dot{x}, \omega t)$ and has a fractal dimension of $d = 2.5$ (Figure 6-8). Using the same data we also saw that we could calculate d from the Poincaré map (Figures 6-9, 6-10). Using the same numerical data from a Runge–Kutta integration, we reconstructed the motion in a pseudo-phase-space using digitized values of $x(t)$ and embedding space dimensions of $m = 2 - 8$. The graphs in Figure 6-13a, b show the correlation function as well as the calculated dimension of the attractor in each embedding space.

One can see in Figure 6-13b that the dimension reaches an asymptote of $d = 2.5$ after $M \sim 4 - 5$, which is in agreement with Taken's theorem.

As an example using experimental data, we describe the work of a group at the French research laboratory at Saclay (e.g., see Malraison et al., 1983; Bergé et al., 1985). They measured the fractal dimension of a convective fluid cell under a thermal gradient (Rayleigh–Benard convection, see Chapter 3). They calculated the fractal dimension using an averaged pointwise dimension (6-2.3) for different sizes of pseudo-phase-spaces. As shown in Figure 6-14, the fractal dimension saturated at a value of $d = 2.8$ when the embedding dimension of the phase space reached 5 or greater. They used 15,000 points and averaged $P_n(\epsilon)$ over 100 random points.

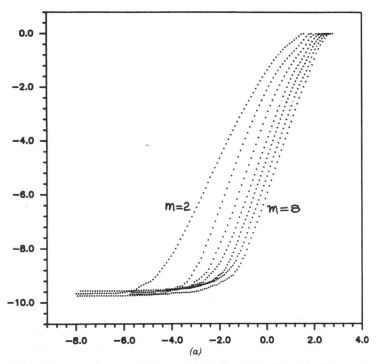

Figure 6-13   (a) log C versus log ϵ for the two-well potential problem for different dimension embedding spaces. Time history data identical to that in Figures 6-8 and 6-10.

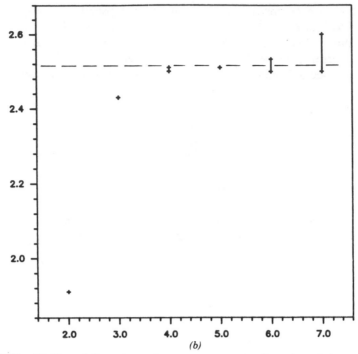

Figure 6-13 (*b*) Fractal dimensions of attractor versus the dimension of the embedding space.

However, they also found regimes of chaotic flow where no clear slope of $\log C(\epsilon)$ versus $\log \epsilon$ existed.

Similar results for the flow between two cylinders (Taylor–Couette flow) has been reported by a group from the Soviet Union (L'vov et al., 1981). They claim to measure the information dimension. Figure 6-15 shows the value of the slope of $\log C(\epsilon)$ versus $\log \epsilon$ as a function of ϵ. This is characteristic of these measurements. The slope values at small ϵ reflect instrumentation noise, while the values at large ϵ are those for which the size of the covering sphere or hypercube reaches the scale of the attractor.

Using such techniques, one can determine how the fractal dimension changes as some control parameter in the experiment is varied. For example, in the case of Taylor–Couette flow (see Figure 3-37), Swinney and coworkers have measures the change in d as a function of the Reynolds number (Figure 6-16; see Swinney, 1985).

In another fluid experiment, Ciliberto and Gollub (1985) have studied chaotic excitation of surface waves in a fluid. The surface wave chaos was excited by a 16 Hz vertical amplitude frequency; 2048 points were sampled with a sampling time of 1.5 s or around 300 orbits. Using the embedding

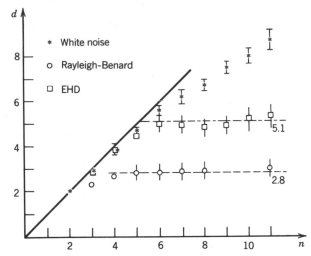

Figure 6-14 Fractal dimension versus dimension of embedding pseudo-phase-space for measurements of electrohydrodynamic fluid flow, Rayleigh–Benard flow (Chapter 3), and white noise [From Malraison et al, 1983).

space technique, they measured both the correlation dimension ($d_c = 2.20 \pm 0.04$) and the information dimension ($d_I = 2.22 \pm 0.04$), both of which reached asymptotic values when the embedding space dimension was 4 or greater. (See also Figure 5-8.)

Holzfuss and Mayer-Kress (1986) have examined the probable errors in estimating dimensions from a time series data set. The three methods studied involved the correlation dimension, averaged pointwise dimension, and the averaged radius method of Termonia and Alexandrowicz (1983). They tested each on a set of 20,000 points from a quasiperiodic motion on a 5 torus, which consists of a time history with 5 incommensurate frequencies. Using the pseudo-phase-space method for embedding dimensions of 2–20, they found that the averaged pointwise dimension had the smallest standard deviation of the three. The average was taken over 20% of the reference points, and curves that did not show scaling behavior over a significant portion of the range of r were rejected.

6.4   OPTICAL MEASUREMENT OF FRACTAL DIMENSION

All the methods for calculating the fractal dimension of strange attractors discussed above require the use of a powerful digital micro or minicomputer. From an experimental point of view, however, it is natural to ask

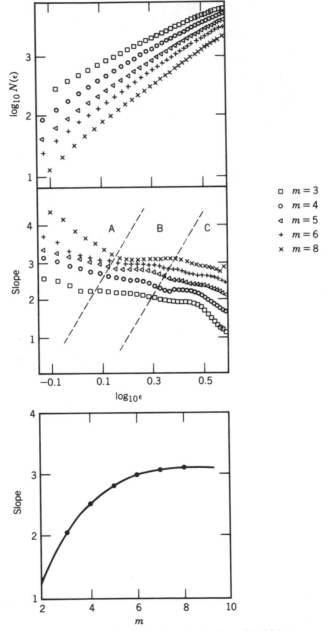

Figure 6-15 Calculation of fractal dimension for chaotic flow of fluid between two rotating cylinders: Taylor–Couette flow (see Chapter 3) [from L'vov et al. (1981) with permission of Elsevier Science Publishers, copyright 1981].

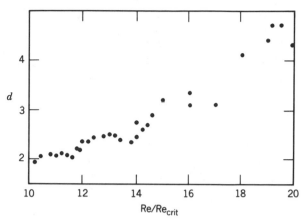

Figure 6-16 Dependence of information dimension on the Reynolds number for flow in a Taylor–Couette system [from Swinney (1985) with permission of Elsevier Science Publishers, copyright 1985].

whether the fractal properties in dynamical systems can be measured directly using *analog devices* in the same way that other dynamical properties such as velocity or acceleration are measured. For general, multidegree-of-freedom systems, the answer is not known; but for simple nonlinear problems, the fractal dimension of a two-dimensional Poincaré map can be measured using optical techniques (Lee and Moon, 1986). This method is based on an optical interpretation of the correlation function (6-2.5).

A diagram illustrating this method is shown in Figure 6-17. We recall that the correlation function involves counting the number of points in a cube or ball surrounding each point in the fractal set of points. The optical method uses a parallel processing feature to perform all the sums at once. Light coming from one film creates a disk of light on another film. If each film is an identical copy of the Poincaré map of the strange attractor, the total light emanating from the second film is proportional to the correlation function. By changing the distance between the two films in Figure 6-17, the radius of the small circle changes and one can obtain the correlation sum as a function of the radius r. A plot of $\log C(r)$ versus $\log r$ then yields the fractal dimension of the Poincaré map D.

If the map is a time triggered Poincaré map, the dimension of the attractor is $1 + D$.

An Optical Parallel Processor for the Correlation Function

A sketch of the experimental setup is shown in Figure 6-18, displaying the optical path of light in this method. The method makes use of two

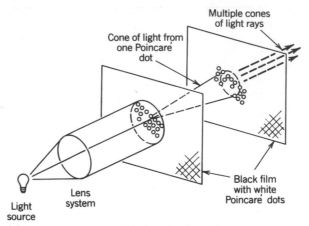

Figure 6-17 Diagram illustrating the parallel processing feature of optically measuring the correlation function and fractal dimension of a planar distribution of points.

properties of classical optics. First, if light is passed through a small aperture of diameter D in the region of Fraunhofer diffraction (if λ is the wavelength, $D \gg \lambda$), then light will cast a circle of radius r, with uniform intensity, on a plane located at a distance L from the aperture. This radius is given by $r = 1.22L\lambda/D$. In our method, the aperture originates from a small dot on the negative of a planar Poincaré map and the small circle of

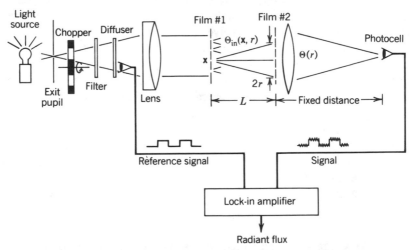

Figure 6-18 Experimental setup of the optical method for measuring fractal dimension [from Lee and Moon (1986) with permission of Elsevier Science Publishers, copyright 1986].

light falls on an identical copy of this negative located at a distance L (Figure 6-18). Second, for incoherent light, the amount of light that emanates from the second negative is proportional to the number of small dots or circles within the circle of illumination. The total amount of light passing through both films is thus proportional to the correlation function $C(r)$. To calculate or vary r, we simply measure or vary L, the distance between the two negatives.

To make these ideas more concrete, let $\Phi(\mathbf{x}, r)$ be the radiant flux behind film #2 due to the flux $\Phi_{in}(\mathbf{x})$ entering the circular aperture at $\mathbf{x}$ on film #1:

$$\Phi(\mathbf{x}, r) = n(x, r) A \frac{\Phi_{in}(\mathbf{x})}{\pi r^2} \tag{6-4.1}$$

where $n(\mathbf{x}, r) = \sum_j H(r - |\mathbf{x} - \mathbf{x}_j|)$ is the number of apertures located within the circle of light illuminated by the flux in the aperture at $\mathbf{x}$, and A is the area of the aperture of a point on film #1. One can see that Φ depends on both n and r explicitly. However, we would like a measure of n alone. Using the linear relation between r and L, we define an adjusted radiant flux $\Phi^* = (r/r_0)^2 \Phi$, where r is the radius of the illuminated area when $L = L_0$ (L_0 is a convenient reference distance). Summing over all points in film #1,

$$\sum_{k=1}^{N} \Phi^*(\mathbf{x}, r) = \left(\frac{r}{r_0}\right)^2 \sum \Phi(\mathbf{x}, r) = \frac{A}{\pi r_0^2} \sum \Phi_{in}(\mathbf{x}) n(r) \tag{6-4.2}$$

When the incident light intensity is uniform over film #1, we find

$$\left(\frac{L}{L_0}\right)^2 \sum_{k=1}^{N} \Phi_0(\mathbf{x}_k, r) \approx \sum_{k=1}^{N} n(r) \approx C(r) \tag{6-4.3}$$

The maps can be obtained from either numerical solution of a third-order system of equations or from experimental data. The light passing through film #2 was focused onto a photocell for the light flux measurement. A light filter (orange-amber color filter) was used at the light source to optimize the photocell response around 6328 Å. The dot size on the negatives was less than 0.2 mm so that $D/\lambda \simeq 300$, which satisfies the Fraunhofer diffraction criterion.

The output voltage from the photocell contained a lot of noise. To extract the signal from the noise, a mechanical light chopper and a lock-in

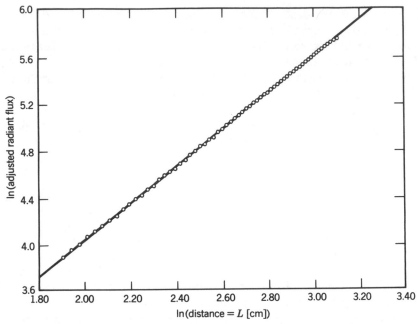

Figure 6-19 Radiant flux versus distance between two films of Poincaré maps on a log–log scale for data from the vibration of a buckled beam [from Lee and Moon (1986) with permission of Elsevier Science Publishers, copyright 1986].

amplifier were used in the signal processing. The chopper was operated at approximately 100 Hz to avoid power line noise.

The radiant flux behind film #2 was measured at the photocell as a function of the distance between films, and the adjusted radiant flux (6-4.2) versus L was plotted on a log–log scale as shown in Figure 6-19. Theoretically, the slope of this curve should give the fractal dimension (6-2.5).

Calculations of fractal dimensions using the correlation function $C(r)$ have shown that there is an optimum range of r to measure the slope. For small r, one encounters the noise error in generating the original map (increasing slope); for large r, one reaches the size of the attractor itself which results in a saturation of $C(r)$ (leading to decreasing slope). A plot of the slope as a function of r is shown in Figure 6-20. One can see that the slope reaches a plateau for a certain range of r or film distance L. This plateau value was chosen as the fractal dimension. The data were obtained from a Runge–Kutta simulation of the forced, two-well potential equation (6-3.7). The 4000 points were generated by taking a Poincaré map synchro-

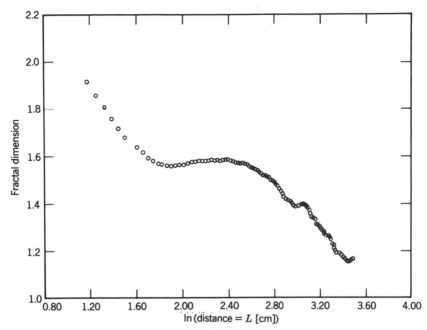

Figure 6-20 Slope versus distance between films L or radius r for data similar to that in Figure 6-19 [from Lee and Moon (1986) with permission of Elsevier Science Publishers, copyright 1986].

nous with the driving frequency. The adjusted radiant flux output was measured at approximately 200 values of L. However, only the linear section of log C versus log L is plotted in Figure 6-19. The slopes in Figure 6-20 were based in 30 points/local averages of the slope of the log $C(r)$ curve.

A comparison of the optically measured fractal dimension with those calculated from the numerical data of Moon and Li (1985a) is shown in Table 6-3 for several values of the damping. The results, as one can see, are remarkably good.

A comparison of the optical and numerical methods for experimental Poincaré maps for the buckled beam is also shown in Table 6-3. In this set of tests, the phase of the Poincaré map trigger was changed. The optical measurement of fractal dimension confirms the results of the numerical method, namely, that the dimension is independent of the phase of the map. This implies that the dimension of the strange attractor itself is $1 + D$, where D is the planar map dimension.

TABLE 6-3 Optically Measured Fractal Dimension for Computer-Simulated and Experimental Poincaré Maps

Numerical Poincaré Map [Eq. (6-3.7)]		
Damping	Calculated[a]	Measured
0.075	1.565[b]	1.558
0.105	1.393	1.417
0.135	1.202	1.162

Experimental Poincaré Map			
Phase Angle	Calculated[a]		Measured
0°	1.741[b]	1.628[c]	1.678
45°	1.751	1.627	1.671
90°	1.742	1.638	1.631
135°	1.748	1.637	1.676
180°	1.730	1.637	1.635

[a] Moon and Li (1985a).
[b] Based on four smallest log r points in log C versus log r.
[c] Based on seven smallest log r points in log C versus log r.

6.5 FRACTAL BASIN BOUNDARIES

Basins of Attraction

In most physical *linear* systems, there is just one possible motion for a given input. For example, the response of a linear mass–spring–damper system to an initial impulse force is just a decaying response, where the mass eventually comes to rest. Such a system has but one attractor, namely, the equilibrium point. However, in nonlinear systems, it is possible for more than one outcome to occur depending on the input parameters such as force level or initial conditions. For example, the system may have more than one equilibrium position or it may have more than one periodic or nonperiodic motion as in certain self-excited systems.

Equilibrium positions, periodic or limit cycle motions, are called *attractors* in the mathematics of dynamical systems. The range of values of certain input or control parameters for which the motion tends toward a given attractor is called a *basin of attraction* in the space of parameters. If there are two or more attractors, the transition from one basin of attraction to another is called a *basin boundary* (see Figure 6-21). In classical prob-

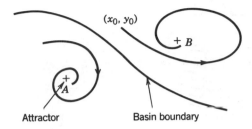

Figure 6-21 Sketch of two dynamic attractors in phase space and the boundary between their basins of attraction in initial condition space.

lems, we expect the basin boundary to be a smooth, continuous line or surface as in Figure 6-21. This implies that when the input parameters are away from the boundary, small uncertainties in the parameters will not affect the outcome. However, it has been discovered that in many nonlinear systems, this boundary is nonsmooth. In fact it is fractal—hence the term *fractal basin boundary*. The existence of fractal basin boundaries has fundamental implications for the behavior of dynamical systems. This is because small uncertainties in initial conditions or other system parameters may lead to uncertainties in the outcome of the system. Thus predictability in such systems is not always possible (see Grebogi et al., 1983b, 1985a, 1986).

Sensitivity to Initial Conditions: Transient Motion in a Two-Well Potential

Before we examine a problem with a fractal basin boundary, it is instructive to look at a case where the basin boundary is smooth, but the outcome is sensitive to initial conditions. This is the case of the *transient* dynamics of a particle with damping. This one degree-of-freedom example is a simple model for the postbuckling behavior of an elastic beam or a particle in a two-well potential. The equation of motion for this problem is

$$\ddot{x} + \gamma\dot{x} - \tfrac{1}{2}\dot{x}(1 - x^2) = 0 \qquad (6\text{-}5.1)$$

Unlike the related problem with periodic forcing, the complete dynamics can be described in a two-dimensional phase plane $(x, y = \dot{x})$. The displacement and time have been normalized such that the two stable equilibrium positions in the phase plane are $(\pm 1, 0)$ and the undamped natural frequency is 1 radian per second. The control parameters are the damping γ and initial conditions $x(0) = x_0$, $\dot{x}(0) = y_0$. While there are three equilibrium positions, $x = 0, \pm 1$, only the latter two are stable so that we will have *two competing basins of attraction*.

Dowell and Pezeshki (1986) have examined the basins of attraction for this problem as illustrated in Figures 5-25 and 6-22. They subdivided the

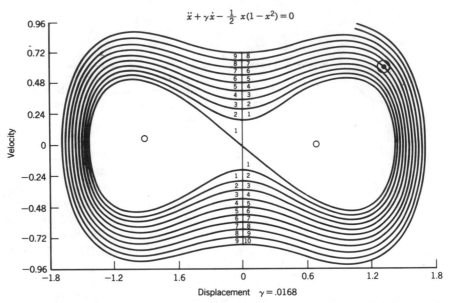

Figure 6-22 Basins of attraction for the unforced, damped motion of a particle in a two-well potential (Dowell and Pezeshki, 1986). The numbers indicate the number of times the trajectory crosses $x = 0$, before going to one of the two equilibrium points at $x = \pm 1$.

basins into how many times the particle orbits cross the $x = 0$ axis before settling down to $x = \pm 1$. One can see that for large initial conditions there are alternating bands where the particle will eventually go to the left or right attractor. Although these boundaries are smooth, the size of the bands approaches zero as the damping $\gamma \to 0$. Thus, if there is some finite uncertainty in the initial conditions as denoted by the circle of radius ϵ in Figure 6-22, one has no certainty of which attractor the particle will go toward if $\epsilon > \epsilon_0(\gamma)$, where $\lim \epsilon_0 \to 0$ as $\gamma \to 0$. For finite damping, we can obtain certainty of the end state only if we have accurate information about the initial state.

In the next example, we show a fractal basin boundary where the outcome is always uncertain no matter how small ϵ is; that is, $\epsilon_0 = 0$.

Fractal Basin Boundary: Forced Motion in a Two-Well Potential

In this section, we examine the periodic forcing of a particle in a two-well potential:

$$\dot{x} = y$$
$$\dot{y} = -\gamma y + \tfrac{1}{2}x(1 - x^2) + f_0 \cos \omega t \qquad (6\text{-}5.2)$$

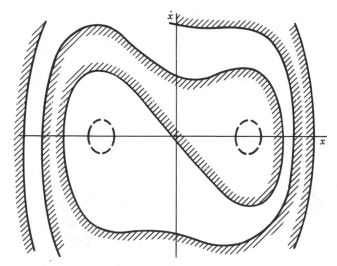

Figure 6-23 Smooth basin boundary for low-amplitude forcing of two-well potential oscillator. The attractors are periodic orbits about left and right equilibrium points [from Moon and Li (1985b) with permission of The American Physical Society, copyright 1985].

As discussed in earlier chapters, the dynamics of the particle can be described in a three-dimensional phase space $(x, y, z = \omega t)$. In the earlier discussions, however, we focused on chaotic motions for this system. Here we only consider motions that are *periodic* about either the left or right equilibrium positions, $x = \pm 1$. Thus, the attractors in this problem may be considered limit cycles. [If we take a Poincaré map of the asymptotic motion, we will have a finite set of points near one of the equilibrium positions $(\pm 1, 0)$.] Here we do not distinguish between period 1 or period 3 subharmonics. We assume that the forcing f_0 is small enough to avoid chaotic vibrations and high-period subharmonics.

In this example, we fix γ, f_0, and ω and vary the initial conditions. The results are shown in Figures 6-23–6-25 and are obtained from numerical simulation using a fourth-order Runge–Kutta integration algorithm (see Moon and Li, 1985b, for details).

The results in Figure 6-23 show that when f_0 is small enough, the basin boundary is smooth, but where f_0 is greater than some critical value, the boundary becomes fractal-like, as shown in Figure 6-24. (This figure is based on the integration of 160×10^3 initial conditions.) To ascertain whether this boundary is fractal, we have taken a small region of initial condition space and have expanded this region. The results are shown in Figure 6-25. Thus, we see that on a finer and finer scale the boundary shows

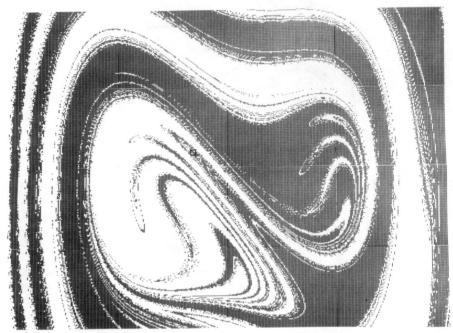

Figure 6-24 Fractal-like basins of attraction for the forced, two-well potential problem for forcing amplitude above the Melnikov criterion (5-3.28) [from Moon and Li (1985b) with permission of The American Physical Society, copyright 1985].

evidence of fractal structure. These results have important implications for classical dynamics insofar as predictability goes.

Homoclinic Orbits: A Criterion for Fractal Basin Boundaries

While the main theme of this book has been chaotic dynamics, the results of the previous section demonstrate that one of the properties of chaotic dynamics, namely, parameter sensitivity and unpredictability, may also be characteristic of certain nonchaotic motions. This prospect stirs terror in the computers of those engineers involved in numerical simulation of nonlinear systems. In such systems, the output of a calculation may be sensitive to small changes in variables such as initial conditions, control parameters, round-off errors, and numerical algorithm time steps. This lack of robustness may exist even when the problem is a transient one or has a periodic output.

It is of great interest for numerical analysts to have a criterion for when a particular nonlinear system will have or not have sensitivity to system

Figure 6-25 Enlargement of a small rectangular region of initial condition space in Figure 6-23 showing fractal-like structure on a finer scale [from Moon and Li (1985b) with permission of The American Physical Society, copyright 1985].

parameters. To date, there is no general test for predictability in nonlinear dynamics, but certain clues are emerging as illustrated by the dynamics of the two-well potential problem.

First, we expect that those systems most susceptible to fractal basin boundary behavior will be those with multiple outcomes, such as multiple equilibrium states or periodic motions. For example, if we consider the impact of an elastic–plastic arch (see Symonds and Yu, 1985; Poddar et al., 1986) or periodic excitation of a rotor or pendulum, there are at least two possible outcomes. In the case of the arch, the end state could be either the arch bend up or down. In the case of the rotor, one could have rotation clockwise or counterclockwise.

The second clue to establishing the possibility of fractal basin boundaries is more subtle and requires more mathematical intuition. We have seen in both Chapters 1 and 5 that nonlinear systems which tend to stretch and fold regions of phase space in what are called *horseshoe maps* have a certain element of sensitivity to initial conditions as well as a variety of sub-

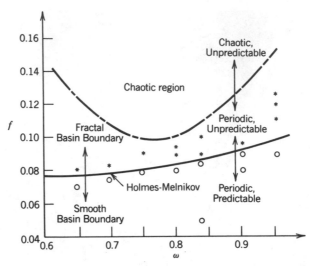

Figure 6-26 Homoclinic orbit criterion (5-3.28) for the two-well potential problem with fractal-like and smooth basin boundary observation from numerical studies [from Moon and Li (1985b) with permission of The American Physical Society, copyright 1985].

harmonic solutions. As discussed in Chapter 5, it was shown that horseshoe map properties result when the Poincaré map associated with the flow in phase space develops homoclinic points in dissipative nonlinear systems. A criterion was derived by Holmes (see Guckenheimer and Holmes, 1983) using a method by Melnikov [Eq. (5-3.20)]. In the case of the forced motion of a particle in a two-well potential, it turns out that this criterion gives a very good indication of fractal basin boundaries even when the motion is *not* chaotic. The criterion for the equation of motion (6-5.2) is given by

$$f_0 > \frac{\gamma\sqrt{2}}{3\pi\omega}\cosh\left(\frac{\pi\omega}{\sqrt{2}}\right) \qquad (6\text{-}5.3)$$

Evidence for this conclusion is given in Figure 6-26 (e.g., see Moon and Li, (1985b). This figure summarizes the results of many calculations of basin boundaries similar to those in Figures 6-23–6-25. Below the Holmes–Melnikov criterion, the numerically calculated basin boundary appears to be smooth, while above the criterion curve, the boundary appears fractal.

The connection between homoclinic orbits and fractal basin boundaries is not entirely a mystery especially if we examine the results in Figure 6-27. In this figure, we have superimposed two calculations. The first is the basin

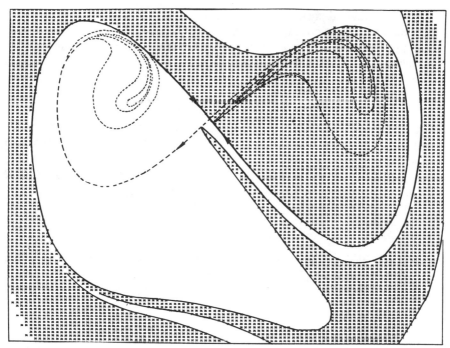

Figure 6-27 Superimposed plots of the basins of attraction of the forced, two-well potential problem and the associated stable and unstable manifolds of the Poincaré map at the critical force level (5-3.28) [from Moon and Li (1985b) with permission of The American Physical Society, copyright 1985].

boundary for the two-well potential for a force amplitude just below the Holmes–Melnikov curve. We can see that the boundary has developed a long finger as compared with that in Figure 6-23 for a smaller force. The second calculation in Figure 6-27 is the determination of the stable and unstable manifolds of the Poincaré map which emanate out of the saddle point near the origin. The first observation is that the basin boundary is *identical* to the stable manifold of the Poincaré map. The second observation is that the unstable manifolds, shown as the dotted curves, are just touching the stable manifolds. This is to be expected since at the criterion the two manifolds touch and form homoclinic points. In theory, beyond this criterion, the two manifolds of the Poincaré map must touch an infinite number of times which results in an infinite folding of the stable manifold and hence an infinite folding of the basin boundary and the resulting fractal properties. These results have yet to be confirmed for other systems but a test of these ideas is under study (e.g., see McDonald et al., 1985).

In any case, there are clues that for many dynamical systems the multiplicity of solutions and the existence of homoclinic orbits or horseshoe map properties may provide a criterion for fractal basin boundaries and predictability in nonlinear systems.

Dimension of Basin Boundaries and Uncertainty

Yorke and coworkers at the University of Maryland have produced numerous studies of basin boundaries, fractals, and chaos. In one study they have shown that the fraction ϕ of uncertain initial conditions in the phase space as a function of the radius of uncertainty ϵ is related to the fractal dimension of the basin boundary (e.g., see McDonald et al., 1985),

$$\phi \approx \epsilon^{D-d}$$

where D is the dimension of the phase space and d is the capacity fractal dimension of the basin boundary. When the boundary is smooth, $d = D - 1$ or

$$\phi \sim \epsilon$$

For example, if the relative uncertainty in initial conditions were $\epsilon = 0.05$, then the uncertainty of the outcome as a fraction of all initial conditions would be $\phi \approx 22\%$ when $d = 1.5$ and $D = 2$.

A technique for calculating d for basin boundaries is described in a number of the Maryland group papers. The technique differs from that for trajectories since the boundary points are never given but are formed from the set of points that lie in neither of two attracting sets. Such fractal sets have been labeled "fat fractals." (See Grebogi et al., 1985c) for a discussion of fat fractals and their application to basin boundary calculations.)

Transient Decay Times: Sensitivity to Initial Conditions

In the preceeding discussion we described how the development of a fractal basin boundary leads to uncertainty about which attractor the system will approach as $t \to \infty$. However, one may also be interested in how much time it takes to approach the attractor. Dowell and Pezeshki (1987) have calculated an initial-condition–transient-time plot for the two-well potential as shown in Figure 6-28. In this diagram each point is coded in color or shade to represent the transient time to approach a periodic orbit around either the left or right potential well. The two wells are not distinguished, only the transient times. They observed fractal-looking patterns when the

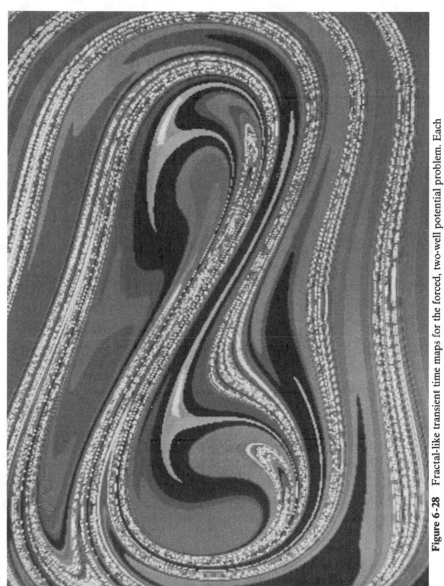

Figure 6-28 Fractal-like transient time maps for the forced, two-well potential problem. Each shade represents a different time for the motion to approach a steady periodic orbit [from Pezeshki and Dowell (1987)].

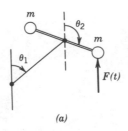

(a)

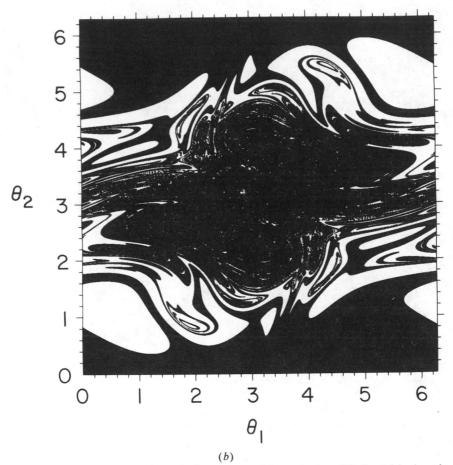

(b)

Figure 6-29 *(a)* Diagram of a pulsed torque two-link mechanism. *(b)* Fractal basins of attraction for the periodically pulsed double rotor. The two attractors are a period-4 motion and an equilibrium solution [from Kostelich et al. (1986) with permission of Elsevier Science Publishers, copyright 1986].

252

forcing amplitude was above the homoclinic orbit criterion (6-5.3). This means that, given some uncertainty in initial conditions, the transient decay time as well as the particular attractor is unpredictable for certain nonlinear problems.

Other Applications

In Chapter 3 we examined the dynamics of an impulsively kicked rotor (3-2.19). The dynamics of a kicked double rotor have been studied by Kostelick et al. (1987). In this two-degree-of-freedom problem (Figure 6-29a) the motion between periodic impulses is linear and can be integrated between kicks. They thus obtain a four-dimension map for which they determine the basins of attractors for two *periodic* attractors. For certain amplitudes of the forcing impulse, the initial condition plot ($\dot{\theta}_1(0) = 0$, $\dot{\theta}_2(0) = 0$) in the (θ_1, θ_2) plane looks highly fractal, as shown in Figure 6-29b.

In another study, Iansiti et al. (1985) and Gwinn and Westervelt (1985) have used the equation of a periodically forced pendulum to model the dynamics of a solid state electronic device known as a Josephson junction. They too obtain a fractal looking basin boundary for different initial conditions.

In a doctoral dissertation at Cornell, Li (1987) has looked at the dynamics of a two-degree-of-freedom system in a four-well potential with periodic forcing. The problem is similar to that described in Chapter 3 (3-3.7) but with four energy wells instead of two. When the forcing amplitude is small, the long time motion is about one of the four equilibrium positions. With the initial velocities equal to zero, one of four colors or shades is given to each initial condition ($x_1(0)$, $x_2(0)$). Li (1987) has shown that the resulting boundary between the four basins of attraction can become fractal as the force level approaches a critical value determined by homoclinic orbit criteria, as shown in Figure 6-30.

Fractal Boundaries for Chaos in Parameter Space

We have seen how small changes in initial conditions can dramatically change the type of output from a dynamical system. It is natural to ask whether a similar sensitivity exists in the other parameters that control the dynamics, such as forcing amplitude or frequency or the damping or resistance in a circuit. One example is discussed here—a fractal experimental boundary between chaotic and periodic motions in a forced, one-degree-of-freedom oscillation.

When two or more types of motion are possible in a system, one usually determines the range of parameters for which one or another type of

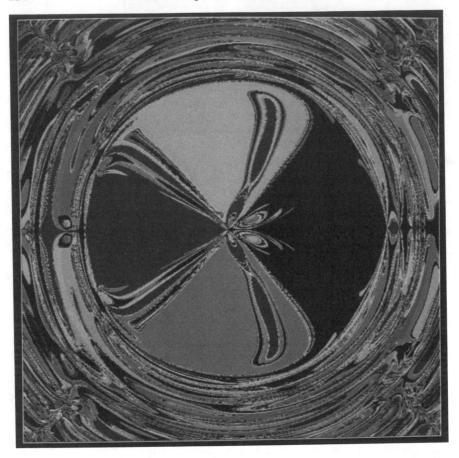

Figure 6-30 Basin boundaries for four periodic attractors showing fractal-like structures. The data is numerically generated for the periodic forcing of a particle in planar motion in a four-well potential. The initial velocities are zero. (From the doctoral dissertation of G. K. Li at Cornell University, 1987.) The axes represent initial positions in the plane.

motion will exist. In the case of the forced motion of a particle in a two-well potential (see Chapters 2 and 5), it is of great interest to know when chaotic motions or periodic motions will occur when the input force is periodic. The equation that describes this oscillation is by now familiar to the reader [Eq. (6-5.2)]. In this problem, we have used a nondimensionalization procedure to eliminate all but three parameters (γ, f, ω). As discussed in Chapter 5, both Holmes (1979) and Moon (1980a) derived criteria relating (γ, f, ω) for when chaotic motion would occur. These relations [Eqs.

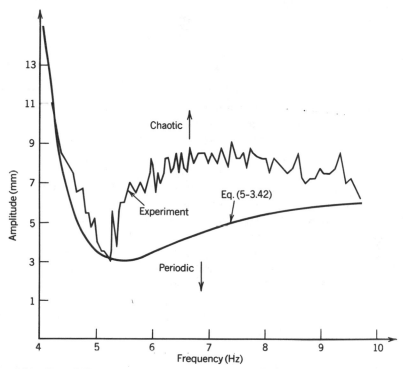

Figure 6-31 Fractal-like boundary between chaotic and periodic motion in the forcing-amplitude–frequency plane. Experimental data from the vibration of a buckled beam [from Moon (1984) with permission of The American Physical Society, copyright 1984].

(5-3.28) and (5-3.42)] have the form

$$f > F(\omega, \gamma) \tag{6-5.4}$$

Fixing the nondimensional damping γ, both criteria are smooth curves in the (f, ω) plane as shown in Figure 6-31. When these criteria are compared with experimental data (see Moon, 1984), however, two differences are obvious. First, the theoretical criteria are lower bounds and second, the experimental criterion looks ragged and may therefore be fractal.

The experiments were carried out on the now familiar buckled, steel, cantilevered beam placed above two permanent magnets (Figure 2-2b). The elastic beam, magnets, and support are placed on an electromagnetic shaker which drives the system at a given amplitude A_0 and frequency ω. The nondimensional force in Eq. (6-5.2) is related to this forcing amplitude by

$$f_0 = -A_0\omega^2$$

The experiments were carried out by fixing the forcing frequency and slowly increasing the driving amplitude of the shaker. With the beam vibrating initially with periodic motion about one of the buckled equilibrium positions, the amplitude was increased until the tip of the beam jumped out of the initial potential well.

To determine whether the motion was chaotic or periodic. Poincaré maps were used. The motion was measured by strain gauges attached to the beam at the clamped end, and the strain versus strain rate served as the phase plane. Poincaré maps of these signals were synchronized at the driving frequency. Chaos was determined when the finite set of points of the Poincaré map (as observed on an oscilloscope; see Chapter 4) became unstable and a Cantor-setlike pattern appeared on the screen.

At least five sets of data for chaotic boundaries were taken for different beam–magnet configurations and all showed a nonsmooth behavior. In the data shown in Figure 6-31, approximately 70 frequencies were sampled between 4 and 9 Hz.

To determine if the boundary between chaotic and periodic motions is fractal, the fractal dimension of the set of experimental points was measured. First, we connected the points with straight line segments. Second, we used the caliper method to measure the length of the boundary as a function of caliper size. This is the same method described by Mandelbrot (1977) to measure the fractal dimension of the coastline of various countries. Thus, we are approximating the experimental boundary by N line segments each of length ϵ. As we decrease the caliper size, ϵ (the number of line segments needed to approximate the curve) increases. The total length is then

$$L = N(\epsilon)\epsilon \qquad\qquad (6\text{-}5.5)$$

For a nonfractal curve, $N \simeq \epsilon^{-1}$ or $N = \lambda/\epsilon$ so that λ becomes a measure of the length of the boundary. However, for fractal curves, such as the Koch curve, $N = \lambda\epsilon^{-D}$, where ϵ is small and D is not an integer. Thus, by measuring L versus ϵ,

$$L = \lambda\epsilon^{1-D} \qquad\qquad (6\text{-}5.6)$$

we can obtain the fractal dimension by measuring the slope of the log L

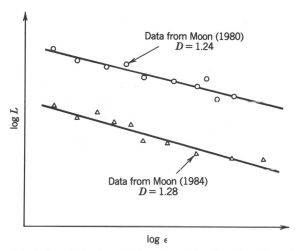

Figure 6-32 Calculation of the fractal dimension of the chaos boundary of Figure 6-31.

versus log ϵ curve, or

$$D = \lim_{\epsilon \to 0} \left(1 + \frac{\log L}{\log(1/\epsilon)} \right) \qquad (6\text{-}5.7)$$

One can show that this procedure is equivalent to the idea of covering the set of points with small squares as discussed in the definition of the capacity fractal dimension [Eq. (6-1.2)].

The results of this series of measurements are shown in Figure 6-32 for two sets of data. The lengths of the boundary curves appear to increase with decrease in caliper size and they imply a fractal dimension of between 1.24 and 1.28. Thus, there is convincing evidence that the boundary curve between periodic and chaotic regimes in the parameter space of (f, ω) is fractal. It should be noted, however, that while the single-mode description of the chaotic elastic beam [Eq. (6-5.2)] agrees very well with the experimental results insofar as Poincaré maps are concerned, the actual experiment has infinitely many degrees of freedom which one hopes do not influence the low-frequency behavior. However, it may be possible that higher modes could influence or are even essential to the fractal nature of the boundary curve in Figure 6-31. Further research on this question is necessary to provide a clear answer.

In any event, these results suggest that a clear-cut criterion for chaos may not be possible. The apparent fractal nature of the criterion boundary may be inherent in many systems and one may have to settle for upper or lower bounds on the chaotic regimes.

6.6   COMPLEX MAPS AND THE MANDELBROT SET

Some readers may have seen the beautiful computer-generated color pictures associated with the name of the Mandelbrot set that have appeared in *Scientific American* (August 1985) or in other publications (e.g., see Peitgen and Richter, 1986). These pictures are examples of fractal basin boundaries and fractal parameter-space boundaries and are based on the two-dimensional map involving the complex variable $z = x + iy$,

$$z_{n+1} = z_n^2 + C \qquad (6\text{-}6.1)$$

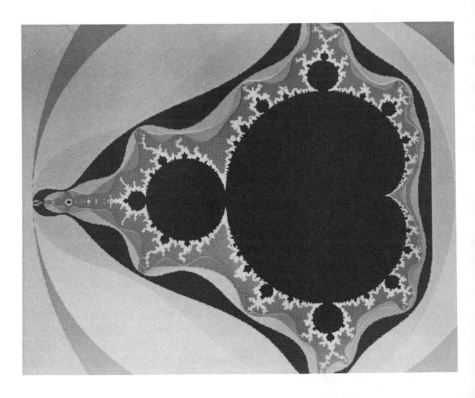

Figure 6-33   Fractal basin of complex parameters for the complex map (6-6.1) for a bounded orbit (sometimes called a Mandelbrot set) [Courtesy of John Hubbard, Cornell University].

where $C = a + ib$ is complex. In terms of real variables, this map becomes

$$x_{n+1} = x_n^2 - y_n^2 + a$$

$$y_{n+1} = 2x_n y_n + b$$

(6-6.2)

In contrast to the work presented in this book, our Poincaré maps are of the form

$$x_{n+1} = f(x_n, y_n)$$

$$y_{n+1} = g(x_n, y_n)$$

(6-6.3)

but $f + ig$ is not an analytic function in the context of the theory of complex variables. Mandelbrot (1977) and others (e.g., see Peitgen and Richter, 1985) have explored the complex C plane for values of C for which the long-term iteration of the map (6-6.1) will remain bounded as $n \to \infty$. The set of points C for which Z_n remains bounded is shown as the black region in Figure 6-33. The boundary of this region has been shown to have fractal properties.

One can also investigate equilibrium or N-periodic points of the map such that

$$z_{n+N} = z_n^2 + C$$

For fixed C, one can study the basin of initial values of (x, y) that leads to specific periodic behavior. This basin boundary has also been found to have fractal properties.

The results of the study of nonlinear difference equations such as (6-6.1) and nonlinear differential equations such as (6-5.2) suggest that the dynamical outcome is sensitive not only to initial conditions, as evidenced by fractal basin boundaries, but also to small changes or uncertainties in the parameters of the problem.

APPENDIX A

Glossary of Terms in Chaotic and Nonlinear Vibrations

Almost periodic: A time history made up of a number of discrete incommensurate frequencies.

Arnold tongues: Refers to the motion of coupled nonlinear oscillators for which the ratio of frequencies may become locked at some value p/q (p,q integers). The tongue refers to the shape of the locked region in some parameter space and the name refers to the Soviet dynamicist V. I. Arnold who discovered them.

Attractor: A set of points or a subspace in phase space toward which a time history approaches after transients die out. For example, equilibrium points or fixed points in maps, limit cycles, or a toroidal surface for quasiperiodic motions, are all classical dynamical attractors.

Baker's map: A transformation of the plane (a mapping from the plane to the plane), which takes a rectangular area, stretches it in one direction, shrinks it in a transverse direction, cuts it in half, and places it back over the original area. Similar to horseshoe map. Repeated iterations of the map transform the original set of points into a fractal structure. Named after the operations of a cook or baker who repeatedly forms and reforms a piece of pie dough.

Basin of attraction: A set of initial conditions in phase space which leads to a particular long-time motion or attractor. Usually this set of points is connected and forms a continuous subspace in phase space. However, the boundary between different basins of attraction may or may not be smooth (see *Fractal basin boundaries*).

Bifurcation: Denotes the change in the type of long-time dynamical motion when some parameter or set of parameters is varied, for example, as when a rod under a compressive load buckles—one equilibrium state changes to two stable equilibrium states.

Cantor set: Formally, a set of points obtained on a unit interval by throwing out the middle third and iterating this operation on the remaining intervals. This operation when carried to the limit leads to a fractal set of points on the line with a dimension between 0 and 1 $(\ln 2/\ln 3)$.

Capacity: One of the many definitions of the fractal dimension of a set of points. The basic idea is to count the minimum number of cubes of size ϵ needed to cover a set of points. If this number behaves as ϵ^{-d} as $\epsilon \to 0$, the exponent d is called the capacity fractal dimension.

Catastrophe theory: In many physical systems, the equilibrium points are derived from a potential function by setting the derivatives of this potential with respect to the generalized coordinates equal to zero. Catastrophe theory has to do with the dependence of the number of equilibrium points on the parameters in the problem such as force loads in elastic systems. Near certain critical values of these parameters, this theory predicts that the number of equilibrium points will change in prescribed ways and that these changes are universal for certain classes of potential functions. The roots of the theory are attributed to the French mathematician René Thom. In engineering mechanics, a special version of the theory was developed independently and deals with the sensitivity of critical loads to imperfections in the structure.

Center manifold: In dynamical systems theory, the motions in the neighborhood of an equilibrium point can be classified according to whether the eigensolutions are stable, unstable, or oscillatory. The subspace of phase space which is spanned by the purely oscillatory solutions is sometimes called the center manifold.

Chaotic: Denotes a type of motion that is sensitive to changes in initial conditions. A motion for which trajectories starting from slightly different initial conditions diverge exponentially. A motion with positive Lyapunov exponent.

Circle map: This is a map or difference equation that maps points on a circle onto the original circle. In the theory of two coupled oscillators, some motions in phase space can be viewed as motion on a toroidal surface. A Poincaré section that intersects the smaller diameter of the torus constitutes a circle map.

Combination tones: (See also *Quasiperiodic vibrations.*) In vibrations and acoustics, frequencies that appear as the sum or difference of two fundamental frequencies. More generally, frequencies of the form ($n\omega_1 + m\omega_2$), where n and m are positive or negative integers.

Deterministic: Refers to a dynamic system whose equations of motion, parameters, and initial conditions are known and are not stochastic or random. However, deterministic systems may have motions that appear random.

Duffing's equation: A second-order differential equation with a cubic non-linearity and harmonic forcing $\ddot{x} + c\dot{x} + bx + ax^3 = f_0 \cos \omega t$. Named after G. Duffing (circa 1918).

Equilibrium point: In a continuous dynamical system, a point in phase space toward which a solution may approach as transients decay ($t \to \infty$). In mechanical systems, this usually means a state of zero acceleration and velocity. For maps, equilibrium points may come in a finite set where the system visits each point in a sequential manner as the map or difference equation is iterated. (Also called a fixed point.)

Ergodic theory: In Hamiltonian mechanics (no dissipation), it refers to the randomlike motions of coupled nonlinear systems of particles and the evolution of collective properties of the total system.

Feigenbaum number: A property of a dynamical system related to the period-doubling sequence. The ratio of successive differences between period-doubling bifurcation parameters approaches the number $4.669\cdots$. This property and the Feigenbaum number have been discovered in many physical systems in the prechaotic regime.

Fixed point: See *Equilibrium point.*

Fractal: A geometric property of a set of points in an n-dimensional space having the quality of self-similarity at different length scales and having a noninteger fractal dimension less than n.

Fractal dimension: The fractal dimension is a quantitative property of a set of points in an n-dimensional space which measures the extent to which the points fill a subspace as the number of points becomes very large. (See *Capacity.*)

Global / local motions: Local motions refer to solutions to dynamical systems that do not wander far from equilibrium points. Global solutions concern motion between and among equilibrium points or solutions that are not confined to a small region of phase space.

Hamiltonian mechanics: Formally, a method to derive the equations of motion of an N-degree-of-freedom dynamical system in terms of $2N$ first-order differential equations (Hamilton, 1805–1865). In practice, a Hamiltonian problem often refers to a nondissipative system in which the forces can be derived from a scalar potential.

Hausdorff dimension: A mathematical definition of fractal properties related to the capacity dimension.

Henon map: A set of two coupled difference equations with one quadratic nonlinearity. When one parameter is set to zero, the equations resemble the logistic or quadratic map. Named after a French astronomer.

Heteroclinic orbit: An orbit in a map or difference equation that occurs when stable and unstable orbits from different saddle points intersect.

Homoclinic orbit: An orbit in a map that occurs when stable and unstable manifolds of a saddle point intersect.

Hopf bifurcation: The emergence of a limit cycle oscillation from an equilibrium state as some system parameter is varied. Named after a mathematician who gave precise conditions for its existence in a dynamical system.

Horseshoe map: A map of the plane onto the plane. Points in the lower half of a rectangular domain are stretched and contracted and mapped into a vertical strip in a section of the left half-plane, while points in the upper half are stretched and contracted and mapped onto a vertical strip in the right half-plane. The process is like transforming a rectangular domain into a horseshoe-shaped set of points, hence the name. Similar to the baker's transformation. Repeated iterations can yield a fractal-like set of points.

Intermittency: A type of chaotic motion in which long time intervals of regular, periodic, or stationary dynamical motion are followed by short bursts of randomlike motion. The time interval between bursts is not fixed but is unpredictable.

Invariant measure: A distribution function that describes the long-time probability of finding the motion of a system in a particular region of phase space.

KAM theory: The initials stand for the theorists Kolmogorov, Arnold, and Moser who developed a theory regarding the existence of periodic or quasiperiodic motions in nonlinear Hamiltonian systems (i.e., systems that have no dissipation and in which the forces can be derived from a potential). This theory states that if small nonlinearities are added to a linear system, the regular motions will continue to exist.

Limit cycle: In the engineering literature, a periodic motion that arises from a self-excited or autonomous system as in aeroelastic flutter or electrical oscillations. In the dynamical systems literature, it also includes forced periodic motions. (See also *Hopf bifurcation*.)

Linear operator: Any mathematical operation (e.g., differentiation, multiplication by a constant) in which the action on the sum of two functions is the sum of the action of the operation on each function. Akin to the principle of superposition.

Lorenz equations: A set of three first-order autonomous differential equations that exhibit chaotic solutions. The equations were derived and studied by E. N. Lorenz of M.I.T. in 1963 as a model for atmospheric convection. This set of equations is one of the principal paradigms for chaotic dynamics.

Lyapunov exponents: Numbers that measure the exponential attraction or separation in time of two adjacent trajectories in phase space with different initial conditions. A positive Lyapunov exponent indicates a chaotic motion in a dynamical system with bounded trajectories. Named after the dynamicist Lyapunov (1857–1918) (in some books spelled Liapunov).

Mandelbrot set: If z is a complex variable, the quadratic map $z \rightarrow z^2 + c$ has more than one attractor. Fixing the initial conditions, one can vary the complex parameter c to determine the basin of attraction as a function of c. The basin boundary that results is fractal and the basin is known as the Mandelbrot set after a mathematician at IBM.

Manifold: A subspace of phase space in which solutions with initial conditions in the manifold stay in the manifold or subspace, under the action of the differential or difference equations.

Map, mapping: A mathematical rule that takes a collection of points in some n-dimensional space and maps them into another set of points. When this rule is iterated, a map is similar to a set of difference equations.

Melnikov function: One theory of chaotic motions focuses on the saddle points of Poincaré maps of continuous phase space flows. Near such points there are subspaces where trajectories are swept into the point (stable manifolds) and subspaces where trajectories are swept away from the point (unstable manifolds). The Melnikov function provides a measure of the distance between these stable and unstable manifolds. One theory contends that chaos is possible when these two manifolds intersect or when the Melnikov function has a simple zero. (Named after a Russian mathematician circa 1962.)

Navier–Stokes equations: A set of three partial differential equations governing the velocity field in the flow of an incompressible, linear, viscous fluid (Navier, 1785–1836; Stokes, 1819–1903).

Noise: In experiments, noise usually denotes the small random background disturbance of either mechanical, thermal, or electrical origin.

Nonlinear: A property of an input–output system or mathematical operation for which the output is not linearly proportional to the input. For example, $y = cx^n$ ($n \neq 1$), or $y = x\,dx/dt$, or $y = c(dx/dt)^2$.

Period-doubling: Refers to a sequence of periodic vibrations in which the period doubles as some parameter in the problem is varied. In the classic model, these frequency-halving bifurcations occur at smaller and smaller intervals of the control parameter. Beyond a critical accumulation parameter value, chaotic vibrations occur. This scenario to chaos has been observed in many physical systems but is not the only route to chaos. (See *Feigenbaum number*.)

Phase space: In mechanics, phase space is an abstract mathematical space whose coordinates are generalized coordinates and generalized momentum. In dynamical systems, governed by a set of first-order evolution equations, the coordinates are the state variables or components of the state vector.

Poincaré section (map): The sequence of points in phase space generated by the penetration of a continuous evolution trajectory through a generalized surface or plane in the space. For a periodically forced, second-order nonlinear oscillator, a Poincaré map can be obtained by stroboscopically observing the position and velocity at a particular phase of the forcing function (H. Poincaré, 1854–1912).

Quasiperiodic: A vibration motion consisting of two or more incommensurate frequencies.

Rayleigh–Benard convection: Circulatory patterns in a fluid produced by a thermal gradient and gravitational forces. The chaos model of Lorenz attempted to simulate some of the dynamics of thermal convection.

Renormalization: A mathematical theory in functional analysis in which properties of some mathematical set of equations at one scale can be related to those at another scale by a suitable change of variables. Developed by the Nobel prize winning physicist K. Wilson (Cornell University). Used in the theory of quadratic maps to derive the Feigenbaum number.

Reynolds number: A nondimensional group in fluid mechanics proportional to a velocity parameter and a characteristic length and inversely proportional to the kinematic viscosity. The transition from laminar to turbulent flow in many fluid problems occurs at a critical value of the Reynolds number (O. Reynolds, 1842–1912).

Rotation number: (Also Winding number.) When a system has two oscillators with frequencies ω_1 and ω_2, the rotation number, based on ω_1, measures the average number of orbits of frequency ω_2 in an orbit of ω_1.

Saddle point: In the geometric theory of ordinary differential equations, an equilibrium point with real eigenvalues with at least one positive and one negative eigenvalue.

Self-similarity: A property of a set of points in which geometric structure on one length scale is similar to that at another length scale. (See also *Fractal, Renormalization.*)

Stochastic process: Often refers to a type of chaotic motion found in conservative or nondissipative dynamical systems.

Strange attractor: Refers to the attracting set in phase space on which chaotic orbits move. An attractor that is not an equilibrium point nor a limit cycle, nor a quasiperiodic attractor. An attractor in phase space with fractal dimension.

Surface of section: See *Poincaré section.*

Symbolic dynamics: Refers to a dynamic model in which not only time is discretized but the state variables take on a finite set of values, for example, $(-1, 0, 1)$. Since the set of values is finite, one is free to use any set of symbols, say (L, C, R). A dynamic trajectory then consists of a sequence of symbols. Related also to cellular autonoma.

Taylor–Couette flow: The flow of fluid between two rotating concentric cylinders.

Torus (invariant): The coupled motion of two undamped oscillators is imagined to take place on the surface of a torus, with circular motion around the small radius representing the oscillatory vibration of one oscillator and motion around the large radius direction representing the other oscillator. If the motion is periodic, a closed helical trajectory will wind around the torus. If the motion is quasiperiodic, the orbit will come close to all points on the torus.

Transient chaos: A term describing motion that looks chaotic during a finite time; that is, it appears to move on the strange attractor, but eventually settles into a periodic or quasiperiodic motion.

Unfolding: In the mathematical theory of stability, a term that describes a set of problems which are close to some idealized problem, as when a small amount of asymmetry is introduced into a problem with symmetry or when small damping is added to a nondissipative dynamical problem. The change in stability or dynamical properties of the idealized problem as some nonidealized terms are added is called an unfolding.

Universal property: A property of a dynamical system that remains unchanged for a certain class of nonlinear problems. For example, the Feigenbaum number relating the sequence of bifurcation parameters in period doubling is the same for a certain class of nonlinear, noninvertible, one-dimensional maps.

Van der Pol equation: A second-order differential equation with linear restoring force and nonlinear damping which exhibits a limit cycle behavior. The classic mathematical paradigm for self-excited oscillations. (Named after B. Van der Pol, circa 1927.)

Winding numbers: see *Rotation number*.

APPENDIX B

Numerical Experiments in Chaos

The spirit of the approach to chaotic vibration in this book has been an empirical one of exploring the range of physical phenomena in which chaotic dynamics play a role. While not all readers will have access to a laboratory or have the inclination to do experiments, most readers have some access to digital computers. Thus, this appendix contains a number of numerical experiments using either a PC or minicomputer in which the reader can explore the dynamics of the now classic paradigms of chaos.

B.1 LOGISTIC EQUATION—PERIOD DOUBLING

Perhaps the easiest problem with which to begin study in the new dynamics is the population growth model or logistic equation

$$x_{n+1} = \lambda x_n (1 - x_n)$$

Period-doubling phenomena were observed by a number of researchers (e.g., see May, 1976) and, of course, Feigenbaum (1978), who discovered the famous parameter scaling laws (see Chapters 1 and 5). Two numerical experiments on a desktop computer are fairly easy to perform. In the first plot, we have x_{n+1} versus x_n with a range of $0 \le x \le 1$. The period-doubling regime is below $\lambda = 3.57$. Start with $\lambda < 3.0$ to see a period-1 orbit. To see the long-term orbit plot the first 30–50 iterates with dots and the later iteration with a different symbol. Of course, one can also plot x_n versus n to see both the transient and steady-state behavior. Chaotic orbits may be found for $3.57 < \lambda \le 4.0$. A period 3 window may be found around $\lambda = 3.83$ (see May, 1976).

The next experiment involves generating a bifurcation diagram. In this picture, the long-term values of the orbit are plotted as a function of the control parameter. Start with some initial condition (e.g., $x_0 = 0.1$) and iterate the map for say 100 steps. Then plot x_n for another 50 steps on the vertical axis with the λ value on the horizontal axis (or visa versa). Take step sizes for λ to be around 0.01 and look at the range $2.5 < \lambda < 4.0$. The diagram should produce the classic pitchfork bifurcation at the period-doubling points. Can you calculate the Feigenbaum number from this experiment?

May (1976) also lists other experiments with one-dimensional maps, for example,

$$x_{n+1} = x_n \exp[\lambda(1 - x_n)]$$

He describes this as a model for growth of a single species population which is regulated by an epidemic disease. Look at the region $2.0 < r < 4.0$. The period-doubling accumulation point and chaos begin at $r = 2.6824$ (May, 1976). This article also lists data for several other computer experiments.

B.2   LORENZ EQUATIONS

A fascinating numerical experiment worth trying is the one in Lorenz's original 1963 paper. In this paper, Lorenz simplifies equations derived by Saltzman (1962) based on the fluid convection equations of mechanics (see Chapter 3). Lorenz acknowledges Saltzman's discovery of nonperiodic solutions of the convection equation. Lorenz chose the now classical parameters to study chaotic motions: $\sigma = 10$, $b = \frac{8}{3}$, $r = 28$ for equations

$$\dot{x} = \sigma(y - x)$$

$$\dot{y} = rx - y - xz$$

$$\dot{z} = -bz + xy$$

His data in Figures 1 and 2 of the 1963 paper may be reproduced by choosing initial conditions $(x, y, z) = (0, 1, 0)$ and a time step of $\Delta t = 0.01$ and projecting the solution on either the $z - x$ or $z - y$ planes.

To derive a one-dimensional map based on this flow, Lorenz chooses to look at *successive maxima* of the variable z which he calls M_n. A plot of M_{n+1} versus M_n reveals a map shaped like a tent. Lorenz then goes on to study a simplified version of the map called a tent map—which is a bilinear version of the logistic equation

$$M_{n+1} = 2M_n \qquad \text{if } M_n < \tfrac{1}{2}$$

$$M_{n+1} = 2 - 2M_n \quad \text{if } M_n > \tfrac{1}{2}$$

B.3   INTERMITTENCY AND THE LORENZ EQUATION

An illustration of intermittency may be seen on the computer by numerically integrating the Lorenz equations with a Runge–Kutta algorithm,

$$\dot{x} = \sigma(y - x)$$

$$\dot{y} = -xz + rx - y$$

$$\dot{z} = xy - bz$$

using the parameters $\sigma = 10$, $b = \tfrac{8}{3}$, and $166 \leq r \leq 167$. For $r = 166$, a periodic time history of say $z(t)$ will be obtained, but for $r = 166.1$ or larger, "bursts" or chaotic noise will appear (e.g., see Manneville and Pomeau, 1980). By measuring the average number of periodic cycles between bursts, N (the laminar phase), one should obtain the scaling

$$N \sim \frac{1}{(r - r_c)^{1/2}}$$

where $r_c = 166.07$.

B.4   HENON ATTRACTOR

An extension of the quadratic map on the line to a map on the plane was proposed by the French astronomer Henon:

$$x_{n+1} = 1 + y_n - ax_n^2$$

$$y_{n+1} = bx_n$$

When $b = 0$, one obtains the logistic map studied by May and Feigen-baum. Values of a and b for which one will get a strange attractor include $a = 1.4$ and $b = 0.3$. Plot this map on the $x - y$ plane with graph limits $-2 \le x \le 2$ and $0.5 \le y \le 0.5$. After obtaining the attractor, rescale your graph to focus on one small area of the attractor. Run the map for a much longer time and look for fine-scale fractal structure. If you have the patience or a fast computer, rescale and run again for an even smaller area of the plane. (See Figures 1-20, 1-22.)

If you have a program to calculate Lyapunov exponents, the reported Lyapunov exponent is $\lambda_1 = 0.2$ and the fractal dimension for this attractor is $d_L = 1.264$. One can also vary a and b to see where the attractor exists and to find period-doubling regions of the (a, b) plane (see Guckenheimer and Holmes, 1983, p. 268; Ott, 1981).

B.5   DUFFING'S EQUATION: UEDA ATTRACTOR

This model for an electric circuit with a nonlinear inductor was discussed in Chapter 3. The equations for the model in first-order form are

$$\dot{x} = y$$

$$\dot{y} = -ky - x^3 + B \cos t$$

Chaotic oscillations were studied quite extensively by Ueda (1979). Use a standard numerical integration algorithm such as a fourth-order Runge–Kutta and examine the case $k = 0.1$, $9.8 \le B \le 13.4$. For $B = 9.8$, one should get a period 3. (Take a Poincaré map when $t = 2\pi n$, $n = 1, 2, \ldots$.) The period-3 motion should bifurcate to chaos around $B = 10$. Beyond $B = 13.3$, the motion should become periodic again with a transient chaos regime. (See Figure 3-13.)

Also, compare the fractal nature of the attractor as damping is decreased for $B = 12.0$ and $k = 0.2, 0.1, 0.05$. Note also that for $k = 0.3$ only a small piece of the attractor remains, while for $k = 0.32$ the motion has become periodic.

B.6   TWO-WELL POTENTIAL DUFFING–HOLMES ATTRACTOR

This example has been discussed throughout the book. Several numerical experiments are worth trying. The nondimensional equations are

$$\dot{x} = y$$

$$\dot{y} = -\delta y + \tfrac{1}{2}x(1 - x^2) + f \cos \omega t$$

(This can be put into a third-order autonomous system by setting $z = \omega t$ and writing $\dot{z} = \omega$.) The factor of $\frac{1}{2}$ makes the small-amplitude natural frequency in each well equal to unity. The criterion for chaos for fixed damping $\delta = 0.15$ and variable f, ω has been discussed in Chapter 5. An interesting region to explore is $\omega = 0.8$, $0.1 \le f \le 0.3$. In this regime, one should go from periodic to chaotic to periodic windows in the chaotic region and out of the chaotic region at $f = 0.3$. Another interesting region is $\delta = 0.15$, $\omega = 0.3$, and $f > 0.2$. In all studies, the reader is encouraged to use a Poincaré map. In using a small desktop computer (PC), one can achieve reasonable computing speeds if the program is run in a compiled form. (See Figure 5-3.)

Another interesting experiment is to fix the parameters, say $f = 0.16$, $\omega = 0.833$, and $\delta = 0.15$, and vary the phase of the Poincaré map; that is, plot (x, y) when $t_n = (2\pi/\omega)n + \varphi_0$ and vary φ_0 from 0 to π. One should see an inversion of the map for $\varphi_0 = 0$, π. Is this related to the symmetry of the equations? (See Figure 4-8.)

B.7   CUBIC MAP (HOLMES)

We have illustrated many of the concepts of chaotic vibrations with the model of the two-well potential attractor. The dynamics are described by a nonlinear second-order differential equation (see Chapters 2 and 3) but an explicit formula for the Poincaré map of this attractor has not been found. Holmes (1979) has suggested a two-dimensional cubic map which has some of the features of a negative stiffness Duffing oscillator

$$x_{n+1} = y_n$$

$$y_{n+1} = -bx_n + dy_n - y_n^3$$

A chaotic attractor may be found near the parameter values $b \approx 0.2$ and $d = 2.77$.

B.8   BOUNCING BALL MAP (STANDARD MAP)

[See Holmes (1982) and Lichtenberg and Leiberman (1983).] As discussed in Chapter 3, a Poincaré map for a ball bouncing on a vibrating table can be obtained exactly in terms of the nondimensional impact velocity v_n and phase of the table motion $\varphi_n = \omega t_n$ (mod 2π):

$$v_{n+1} = (1 - \epsilon)v_n + K \sin \varphi_n \qquad \varphi_n \ (\text{mod } 2\pi)$$

$$\varphi_{n+1} = \varphi_n + v_{n+1}$$

where ϵ represents energy lost during impact.

Case 1: $\epsilon = 0$ Conservative Chaos. This case is studied in Lichtenberg and Leiberman (1983) as a model for acceleration of electrons in electromagnetic fields. Iterate the map and plot points on the (v_n, φ_n) plane. To obtain φ (mod 2π) one can use

$$\frac{\varphi}{2\pi} - ABS\left(\frac{\varphi}{2\pi}\right)$$

in advanced BASIC. To get a good picture, you must vary the initial conditions. For example, choose $\varphi = 0.1$ and run the map for several hundred iterations for different values of v between $-\pi < v < \pi$.

The interesting cases are for $0 < K < 1.5$. For $K \ll 1$, one can see quasiperiodic closed orbits around the periodic fixed points of the map. For $K \approx 1$, one should see regions of conservative chaos near the separatrix points. (See Figure 5-21.)

Case 2: $0 < \epsilon < 1$. This case corresponds to a dissipative map where energy is lost at each impact. First try $K \approx 1.2$ and $\epsilon \approx 0.1$. Note that although the early iterations look chaotic, as in Case 1, the motion settles into a periodic orbit. To get fractal-like chaos, raise K to ~ 5.8–6.9. To get a more fractal-looking strange attractor use $\epsilon \approx 0.3$–0.4 and $K \approx 6.0$.

B.9   CIRCLE MAP: MODE LOCKING, WINDING NUMBERS, AND FAIREY TREES

A point moving on the surface of a torus is a conceptual model for the dynamics of two coupled oscillators. The amplitudes of each motion are represented by minor and major radii of the torus and are often assumed to be fixed. The phases of each oscillator are represented by two angles describing positions along the major and minor circumferences. A Poincaré section of the minor circumference of the torus produces a one-dimensional difference equation called the *circle map*:

$$\varphi_{n+1} = \varphi_n + f(\varphi_n)$$

where $f(\varphi)$ is a periodic function.

Each iteration of the map represents an orbit of one oscillator around the large circumference of the torus. A popular example for study is the so-called *standard circle map* (normalized by 2π):

$$x_{n+1} = x_n + \Omega - \frac{K}{2\pi}\sin 2\pi x_n \ (\text{mod } 1)$$

Possible motions observed in this map are periodic, quasiperiodic, and chaotic. To see periodic cycles, plot the points on a circle with rectangular coordinates $u_n = \cos 2\pi x_n$ and $v_n = \sin 2\pi x_n$.

When $K = 0$, Ω represents a *winding number*—the ratio of two frequencies of the uncoupled oscillators. When $K \neq 0$, the map may be periodic when Ω is an irrational number. The oscillators are then said to be *mode locked*. When $0 < K < 1$, one can observe mode-locked or periodic motions in finite-width regions of the Ω axis, $0 < \Omega_k \leq \Omega \leq \Omega_{k+1} < 1$, which of course includes nonrational values of Ω. For example, for $K = 0.8$, a two-cycle can be found for $0.48 < \Omega < 0.52$ and a three-cycle in the region $0.65 < \Omega < 0.66$. To find these regions for $0.7 < K < 1.0$, calculate the winding number W as a function of Ω, $0 \leq \Omega \leq 1$. The winding number is calculated by suspending the mod 1 action and using

$$W = \lim_{N \to \infty} \frac{x_N - x_0}{N}$$

In practice, one has to choose $N > 500$ to get good data. Plotting W versus Ω, one should see a series of plateaus of mode-locked regions. To see more mode-locked regions, one should choose a small $\Delta\Omega$ region and plot W for many points in this small domain.

Each mode-locked plateau in the $W(\Omega)$ plot corresponds to a p/q rational number representing p cycles of one oscillator to q cycles of the other. These p/q ratios are ordered in a sequence called a *Fairey tree*. Given two mode-locked regions r/s and p/q at Ω_1 and Ω_2, respectively, a new mode-locked region will exist somewhere between these two, $\Omega_1 < \Omega < \Omega_2$, with a winding number given by

$$W = \frac{r + p}{s + q}$$

Starting from $0/1$ at $\Omega = 0$ and $1/1$ at $\Omega = 1$, one can begin to generate the whole infinite sequence of mode-locked regions. Most, however, are very narrow. Note that the size $\Delta\Omega$ of these mode-locked regions approaches zero as $K \to 0$ and gets wider as $K \to 1$. The shapes of mode-locked regions in the $K - \Omega$ plane are sometimes called *Arnold tongues*.

B.10 RÖSSLER ATTRACTOR: CHEMICAL REACTIONS, RETURN MAPS

Thus far, each of the principal fields of classical physics has developed a simple paradigm for chaotic dynamics: fluid mechanics—Lorenz equations, structural mechanics—Duffing–Holmes two-well attractor, electrical science—Duffing–Ueda attractor. Another simple model motivated by the dynamics of chemical reactions in a stirred tank is the following proposed by Rössler (1976a):

$$\dot{x} = -(y + z)$$

$$\dot{y} = x + \alpha y$$

$$\dot{z} = \alpha + z(x - \mu)$$

The system often studied is the case $\alpha = \frac{1}{5}$. Period-1, -2, and -4 motions may be found for $\mu = 2.6$, 3.5, and 4.1. Chaotic motions may be found for $\mu > 4.23$.

This model has the properties of a linear oscillator with negative damping and feedback,

$$\ddot{y} - \alpha \dot{y} + y = -z$$

This example is also illustrative of higher-dimensional systems whose dynamics are approximated by a one-dimensional map. Take the Poincaré section for $y = 0$ and plot the x_n values on the $x - z$ plane in the form of a one-dimensional map, that is, x_{n+1} versus x_n. Note the resemblance to the quadratic or logistic map. It should be no surprise that period doubling is observed in this system.

B.11   FRACTAL BASIN BOUNDARIES: KAPLAN–YORKE MAP

An example of a two-dimensional map with a fractal basin boundary is one studied by Kaplan and Yorke (1978) and McDonald et al. (1985):

$$x_{n+1} = \lambda_x x_n \ (\mathrm{mod}\ 1)$$

$$y_{n+1} = \lambda_y y_n + \cos 2\pi x_n$$

where λ_x is an integer. When $\lambda_x = 2$ and $|\lambda_y| < 1$, this map possesses a strange attractor which the reader can explore on the computer. [For $\lambda_y = 0.2$, the fractal dimension is reported to be 1.43 (Russel et al., 1980).]

However, to look at a simple numerical experiment in basin boundaries try the case $\lambda_x = 3$, $\lambda_y = 1.5$. For this case, there are two attractors $y \pm \infty$. Set the scale for $0 \le x \le 1$, $-2.0 \le y \le 2.0$. To get the boundary, choose some initial x value and scan a set of initial y values. For each set of initial conditions, iterate the map until $|y| > 10$ or some other large value. If $y \to +\infty$, leave a blank at the (x_0, y_0) point; if $y \to -\infty$, print a dot. If one scans y_0 from bottom to top once the boundary is crossed, one can omit further y values and choose another x_0.

This example is one of the few for which an explicit formula for the boundary may be obtained:

$$y = -\sum_{j=1}^{\infty} \lambda_y^{-j} \cos\left(2\pi \lambda_x^{j-1} x\right)$$

(see McDonald et al., 1985).

The capacity dimension of this boundary derived by McDonald et al. is $d = 2 - (\ln \lambda_y / \ln \lambda_x) = 1.63 \cdots$. This boundary is continuous, of infinite length, and nowhere differentiable.

B.12 TORUS MAPS

The motion of coupled nonlinear oscillators is sometimes imagined to occur on the surface of a torus. When the number of oscillators is two, a Poincaré section of the torus yields a circle map. However, when the number of oscillators is *three*, the dynamic interaction of the phase of each oscillator takes place on some abstract torus. The Poincaré section of this three-torus yields a two-dimensional map on a two-torus. Grebogi et al. (1985a) have studied such maps and have produced *beautiful* pictures of chaotic attractors. The set of equations takes the form

$$\theta_{n+1} = \left[\theta_n + \omega_1 + \frac{\epsilon}{2\pi} P_1(\theta_n, \psi_n)\right] (\text{mod } 1)$$

$$\psi_{n+1} = \left[\psi_n + \omega_2 + \frac{\epsilon}{2\pi} P_2(\theta_n, \psi_n)\right] (\text{mod } 1)$$

The functions P_1 and P_2 are periodic functions of the form

$$P_\sigma(\theta, \psi) = \sum_{r,s} A_{r,s}^{(\sigma)} \sin\left[2\pi\left(r\theta + s\psi + B_{r,s}^{(\sigma)}\right)\right]$$

where $\sigma = 1, 2$ and (r, s) take on combinations of $(1,0)$, $(0,1)$, $(1,1)$, and $(1, -1)$. The values of $A_{r,s}^{(1)}$, $A_{r,s}^{(2)}$, $B_{r,s}^{(1)}$, and $B_{r,s}^{(2)}$ were chosen randomly by Grebogi et al. The details are listed in their paper in Table 1. Iterations of this map produce spectacular pictures of this strange attractor in the torus, which with high resolution are suitable for framing (see Figures 7, 9, 10, 11 of Grebogi et al., 1985a).

Maps of this kind are also related to the Newhouse–Ruelle–Takens theory of the quasiperiodicity route to chaos.

APPENDIX C

Chaotic Toys

During many lectures given on chaos, I have demonstrated chaotic vibrations with a simple, inexpensive vibrating beam. This chaotic toy has many times made a believer out of a doubting Thomas and has provided motivation to study the often difficult mathematical theory behind chaotic phenomena. In this appendix, I describe several chaotic toys or desktop experiments and also provide some detailed description of the buckled beam experiment (which has been mentioned many times in this book) for the more serious experimenter. During one of these lectures, a physicist (with tongue in cheek) dubbed this experiment the "chaotic Moon-beam." This experiment has had great success in providing both qualitative and quantitative verification of many of the theoretical ideas about chaos.

Another chaotic toy is a version of the forced pendulum sometimes seen in adult toy shops under names such as the "Space ball." A description of this experiment is also given.

Finally, a short description of a simple neon bulb circuit with chaotic flashing lights is provided. For those interested in a simple circuit that exhibits period-doubling phenomena, I recommend the circuit described in Matsumoto et al. (1985) called the "double scroll attractor," which was discovered by L. Chua of University of California at Berkeley. This circuit is described in Chapter 3.

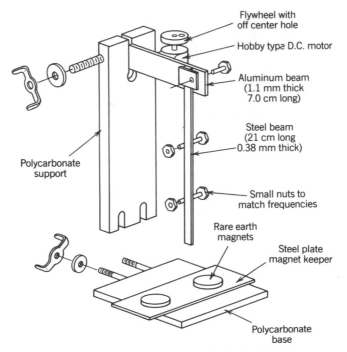

Figure C-1   Exploded view of chaotic elastica toy.

C.1   THE CHAOTIC ELASTICA: A DESKTOP CHAOTIC
VIBRATION EXPERIMENT

This mechanical toy is very inexpensive to build and can demonstrate three different chaotic phenomena:

1.   The two-well potential attractor (or buckled beam)
2.   Bilinear oscillator
3.   Out-of-plane chaotic vibration of a thin beam

A sketch of the device is shown in Figure C-1 for the buckled beam problem. It consists of a small, hobby-type, battery-run motor with an eccentric weight as a source of forced vibrations and a thin steel canti-levered beam with two magnets near the free end of the beam to provide nonlinear buckling forces. Two masses are attached to the thin beam to match dynamically the driving frequency (4–8 Hz) to one or two of the natural modes of the beam. A strong polycarbonate plastic acts as a

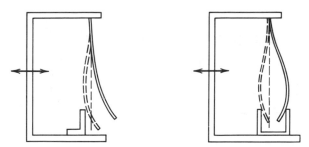

Figure C-2 Elastica and boundary conditions for bilinear beam chaos.

supporting frame and the base can be secured to a table or desktop with double-sided poster adhesive pads. The whole device can be disassembled and carried in a thin box to fit in one's briefcase for travel.

The device works as follows. With a low voltage applied by two or three D-cell batteries across a potentiometer, the aluminum beam at the top is excited by the rotation of the eccentric weight on the motor. With two rare earth magnets below the beam, the beam will vibrate periodically about one of the two stable equilibrium positions (of course, one can use more than two magnets). With the two masses attached, the steel beam will resonate near the second mode so that the beam tip undergoes large deflection. As the motor speed is increased further, the beam will jump from one equilibrium position to the other. Under the right conditions (e.g., magnet spacing, motor speed, mass positions), which usually take about 5 minutes to search for, the beam will perform chaotic motions.

To achieve a more theatrical effect, I have glued a small mirror on the beam and projected a laser beam on a wall or ceiling with spectacular effects as the motion makes the transition from periodic to chaotic motion.

If the magnets are replaced by a thin metallic channel as shown in Figure C-2, one can demonstrate chaotic vibrations of a beam with nonlinear boundary conditions (see Chapter 3). If the metal end constraint is very thin, the audience can hear the nonperiodic or periodic tapping of the beam against the constraints.

C.2 THE "MOON BEAM" OR CHAOTIC BUCKLING EXPERIMENT

As described in Chapters 2, 3, and 4, the forced motion of a buckled cantilevered beam in the field of two strong magnets can be described quite

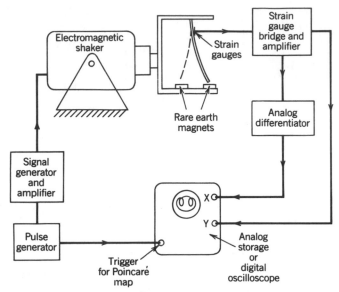

Figure C-3 Professional model of chaotic elastic beam experiment using an electromagnetic shaker.

adequately by a nonlinear differential equation of the Duffing type:

$$\ddot{x} + \dot{x} - \alpha x + \beta x^3 = f \cos \omega t$$

Successful experiments in chaotic vibration have been carried out with two different beams by shaking the beam clamp and magnets with an electromagnetic shaker as shown in Figure C-3. Standard electromagnetic shakers can cost from $2000 to $4000 in 1986 prices. However, the resourceful experimenter can improvise one from a $200 audiospeaker by using the magnet and drive coil in the speaker.

A list of specifications for two elastic beams is given in Table C-1. The best magnets to use are rare earth permanent magnets of 2.5 cm circular diameter.

With this setup, one can obtain Poincaré maps of chaotic motions (Chapter 4), measure the critical force for chaotic motion as a function of frequency (Chapter 5), or measure the fractal dimension of the motion using time series data (Chapter 6).

To obtain electrical signals proportional to the motion, we used two strain gauges glued near the clamped end of the beam. One gauge is placed

TABLE C-1   Parameters for Chaotic Buckled Beam Experiments[a]

	Model A	Model B
Dimensions of elastic steel beam		
Length	18.8 cm (7.4 in.)	12.4 cm (4.9 in.)
Width	9.5 mm (3/8 in.)	12.7 mm (0.5 in.)
Thickness	0.23 mm (0.009 in.)	0.38 mm (0.015 in.)
Constrained layer (for damping)	0.05 mm (0.002 in.)	0.025 mm (0.001 in.)
Natural frequencies		
Cantilevered—no magnets	4.6, 26.6, 73.6 Hz	18 Hz
Cantilevered—with magnets	9.3 Hz	38 Hz
Damping		
Without constrained layer	0.0033	
With steel shim		
and double sided tape layer	0.017	
Buckled displacement with magnets	± 20 mm	± 15 mm
Frequency range for test data	6–12 Hz	21–35 Hz
Driving amplitude range	± 2–5 mm	± 5 mm
Magnets, rare earth, 2.5 cm diameter, 0.2 tesla on one face		

[a] See also Moon (1980a).

on each side and the two resistors (i.e., the gages) are connected to two legs of a Wheatstone bridge.

The output of the bridge is amplified and electronically filtered. One can also design an inexpensive circuit to differentiate the filtered signal. In both devices, one should use care to achieve minimum distortion in both amplitude and phase for frequencies up to at least twice the natural frequency of the beam near one of the two magnets.

Damping is a critical property in this experiment. Most metallic structures have very low damping and the Poincaré maps will look more like Hamiltonian or conservative chaos than the fractal or dissipative chaos. In our experiments, we used constrained layer damping to increase dissipation. A simple way to do this is to put double-sided sticky cellophane tape along the beam and to put a thin shim-type metal layer (0.1 mm) on top of this. When constrained layer damping is placed on each side of the beam, a significant increase in damping can be achieved and some very beautiful fractal-looking Poincaré maps can be obtained.

The reader should see Chapter 4 for other suggestions about experiments in chaotic vibrations.

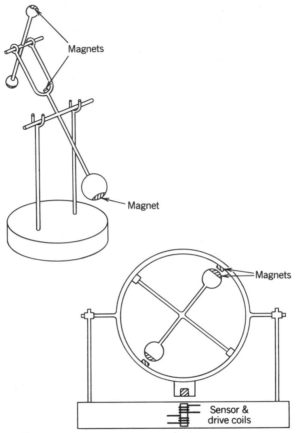

Figure C-4 Double pendulum and "Space ball" chaotic toys.

C.3 A CHAOTIC DOUBLE PENDULUM OR "SPACE BALL"

This toy has several variations, two of which are shown in Figure C-4. The commercial versions are well made (from Taiwan) but I could not find the name of any manufacturer (nor for that matter any patent numbers) on the devices. The basic principles involve the forced motion of a pendulum that interacts with a magnetic circuit in the base. Attached to the primary pendulum is another rotating arm. Several configurations are possible as shown in Figure C-4. In all caes, the pivot point of the second arm is forced by the motion of the driven pendulum. In some versions of this toy, small magnets on both arms interact when the second arm rotates past the primary arm.

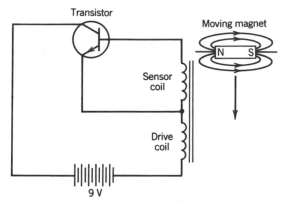

Figure C-5 Pulsed torque circuit for the chaotic pendulum toy.

A simple but clever driving circuit is used to provide current impulses to a driving magnet as shown in Figure C-5. When the lower pendulum oscillates, the magnetic field in the attached magnet generates a voltage in a coil in the base circuit. This voltage is applied to a transistor which begins to conduct when this motion-induced voltage reaches a critical value. During the conduction phase, current can flow out of the 9 V battery into a second coil wrapped around the magnet, thus providing a pulsed torque to the pendulum. In most cases, the motion of the driven pendulum is almost periodic, while the second arm performs chaotic rotations. Professor Alan Wolf of Cooper Union, New York City, and colleagues have analyzed this toy and have shown the motion to be chaotic (the published version was not available at the time of this writing.)

C.4 NEON BULB CHAOTIC TOY

For those with a more electrical bent, my final example is one described by Rolf Landauer of IBM, Yorktown Heights, New York, in 1977 in an internal IBM memorandum entitled "Poor Man's Chaos." A similar study was published by Gollub, Brunner, and Danly (1978). The circuit is shown in Figure C-6 and consists of two neon bulb circuits coupled together. A single circuit can perform relaxation oscillations (e.g., see Minorsky, 1962). When coupled together, the two circuits can exhibit stationary, periodic, or chaotic dynamics that are made visible to the observer by the flashing neon bulb. In the Gollub et al. paper, however, tunnel diodes were used in place of the neon bulbs and inductances were added to the circuits.

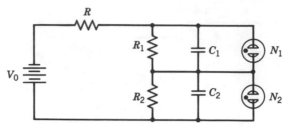

Figure C-6   Circuit for neon bulb experiment [after R. Landauer (1977)].

Landauer was inspired to build this chaotic toy by memories of a similar exercise in a U.S. Navy electronic technician training program in 1945. This gives further evidence to my claim that chaotic vibrations were observed in the past but were seen as curiosities since there were no theoretical foundations for their serious study.

REFERENCES

Abraham, R. H., and Shaw, C. D. (1983). *Dynamics: The Geometry of Behavior*, Aerial Press, Santa Cruz, CA.

Abraham, N. B., Albano, A. M., Das, B., DeGuzman, G., Yang, S., Gioggia, R. S., Puccioni, G. P., and Tredicce, J. R. (1986). "Calculating the Dimension of Attractors From Small Data Sets," *Phys. Lett.* **114A**(5), 217–221.

Arnold, V. I. (1978). *Ordinary Differential Equations*, MIT Press, Cambridge, MA.

Babitsky, V. I., Landa, P. S., Ol'khovoi, A. F., and Perminov, S. M. (1982). "Stochastic Behavior of Auto-Oscillating Systems with Inertial Self-Excitation," *Z. angew. Math. Mech.* **66**(2), 73–81.

Baillieul, J., and Brockett, R. W., and Washburn, R. B. (1980). "Chaotic Motion in Nonlinear Feedback Systems," *IEEE Trans. on Circuits and Systems* **CAS-27**, (11), 990–997.

Baker, N. H., Moore, P. W., and Spiegel, E. A. (1971). "Aperiodic Behavior of a Nonlinear Oscillator," *Q. J. Mech. Appl. Math.*, **24**(4), 391–422.

Bau, H. H., and Torrance, K. E. (1981). "On the Stability and Flow Reversal of an Asymmetrically Heated Open Convection Loop," *J. Fluid Mech.* **106**, 412–433.

Benettin, G., Galgani, L., Giogilli, A., and Strelcyn, J. M. (1980). "Lyapunov Characteristic Exponents for Smooth Dynamical Systems and for Hamiltonian Systems; A Method for Computing All of Them. Part 2: Numerical Application," *Meccanica* **15**, 21–30.

Bergé, P. (1982). "Study of the Phase Space Diagrams Through Experimental Poincaré Sections in Prechaotic and Chaotic Regimes," *Phys. Scr.* **T1**, 71–72.

Bergé, P., Dubois, M., Manneville, P., and Pomeau, P. (1980). "Intermittency in Rayleigh–Benard Convection," *J. Phys. (Paris) Lett.* **41**(15), L341–L345.

Bergé, P., Pomeau, Y., and Vidal, Ch. (1985). *L'Ordre dans le Chaos*, Hermann, Paris.

Brandstater, A., Swift, J., Swinney, H. L., Wolf, A., Farmer, J. O., Jen, E., and Crutchfield, J. P. (1983). "Low-Dimensional Chaos in a Hydrodynamics System," *Phys. Rev. Lett.* **51**(6), 1442–1445.

Brandstater, A., Swift, J., Swinney, H. L., and Wolf, A. (1984). "A Strange Attractor in a Couette–Taylor Experiment," in *Turbulence and Chaotic Phenomena in Fluids*, Proceedings of the IUTAM Symposium (Kyoto), T. Tatsumi (ed.), North-Holland, Amsterdam.

Brillouin, L. (1964), *Scientific Uncertainty and Information*, Academic Press, N.Y.

Brockett, R. W. (1982). "On Conditions Leading to Chaos in Feedback Systems," *IEEE Proc. 21st Conf. Decision Control*, 932–936.

Brorson, S. D., Dewey, D., and Linsay, P. S. (1983). "The Self-Replicating Attractor of a Driven Semiconductor Oscillator," *Phys. Rev. A. Rapid Comm.*

Bryant, P., and Jeffries, C. (1984a). "Bifurcations of a Forced Magnetic Oscillator Near Points of Resonance," *Phys. Rev. Lett.* **53**(3), 250–253.

Bryant, P., and Jeffries, C. (1984b). "Experimental Study of Driven Nonlinear Oscillator Exhibiting Hopf Bifurcations, Strong Resonances, Homoclinic Bifurcations and Chaotic Behavior," Lawrence Berkeley Laboratory report, LBL-16949, January.

Bucko, M. F., Douglass, D. H., and Frutch, H. H. (1984). "Bounded Regions of Chaotic Behavior in the Control Parameter Space of a Driven Non-linear Resonator," *Phys. Lett.* **104**(8), 388–390.

Campbell, D. K., and Rose, H. A. (eds.) (1982). "Order in Chaos," *Proceedings of a Conference Held at Los Alamos National Lab*, May, North-Holland, Amsterdam.

Chirikov, B. V. (1979). "A Universal Instability of Many-Dimensioanl Oscillator Systems," *Phys. Rep.* **52**, 265.

Ciliberto, S., and Gollub, J. P. (1985). "Chaotic Mode Competition in Parametrically Forced Surface Waves," *J. Fluid Mech.* **158**, 381–398

Clemens, H., and Wauer, J. (1981). "Free and Forced Vibrations of a Snap-Through Oscillator," Report of the Institute für Technische Mechanik, Universität Karlsruhe.

Creveling, H. F., dePaz, J. F., Baladi, J. Y., and Schoenhals, R. J. (1975). "Stability Characteristics of a Single-Phase Free Convection Loop," *J. Fluid Mech.* **67**, 65–84.

Croquette, V., and Poitou, C. (1981). "Cascade of Period Doubling Bifurcations and Large Stochasticity in the Motions of a Compass," *J. Phys. (Paris) Lett.* **42**, L537–L539.

Crutchfield, J. P., and Packard, N. H. (1982). "Symbolic Dynamics of One-Dimensional Maps: Entropies, Finite Precursor, and Noise," *Int. J. Theor. Phys.* **21**(6/7), 433–465.

Cvitanovic, P., and Predrag, P. (1984). *Universality in Chaos*, Heyden, Philadelphia, PA.

Dowell, E. H. (1975). *Aeroelasticity of Plates and Shells*, Noordhoff International, Groningen.

Dowell, E. H. (1982). "Flutter of a Buckled Plate as an Example of Chaotic Motion of a Deterministic Autonomous System," *J. Sound Vib.* **85**(3), 333–344.

Dowell, E. H. (1984). "Observation and Evolution of Chaos for an Autonomous System," *J. Appl. Mech.*, paper no. 84-WA/APM-15.

Dowell, E. H., and Pezeshki, C. (1986). "On the Understanding of Chaos in Duffing's Equation Including a Comparison with Experiment," *J. Appl. Mech.* **53**(1), 5–9.

Dubois, M., Bergé, P., and Croquette, V. (1982). "Study of Nonsteady Convective Regimes Using Poincaré Sections," *J. Phys. (Paris) Lett.* **43**, L295–L298.

Eckmann, J. P. (1981). "Roads to Turbulence in Dissipative Dynamical Systems," *Rev. Mod. Phys.* **53**(4), part 1, 643–654.

Evenson, D. A. (1967). *Nonlinear Flexural Vibrations of Thin-Walled Circular Cylinders*. NASA Technical Note, NASA TN D-4090, August.

Everson, R. M. (1986). "Chaotic Dynamics of a Bouncing Ball," *Physica* **19D**, 355–383.

Farmer, J. D., Ott, E., and Yorke, J. A. (1983). "The Dimension of Chaotic Attractors," *Physica* **7D**, 153–170.

Feigenbaum, M. J. (1978). "Qualitative Universality for a Class of Nonlinear Transformations," *J. Stat. Phys.* **19**(1), 25–52.

Feigenbaum, M. J. (1980). "Universal Behavior in Nonlinear Systems," *Los Alamos Sci.* (Summer), 4–27.

Fung, Y. C. (1958). "On Two-dimensional Panel Flutter," *J. Aero/Space Sci.* **25**(3), 145–159.

Glass, L., Guevau, X., and Shrier, A. (1983). "Bifurcation and Chaos in Periodically Stimulated Cardiac Oscillator", *Physica* **7D** (1983), 89–101.

Goldberger, A. L., Bhargava, V., West, B. J., and Mandell, A. J. (1986). "Some Observations on the Question: Is Ventricular Fibrillation 'Chaos'?," *Physica* **19D**, 282–289.

Gollub, J. P., and Benson, S. V. (1980). "Many Routes to Turbulent Convection," *J. Fluid Mech.* **100**(3), 449–470.

Gollub, J. P., Brunner, T. O., and Danly, B. G. (1978). "Periodicity and Chaos in Coupled Nonlinear Oscillators," *Science* **200**, 48–50.

Gollub, J. P., Romer, E. J., and Socalar, J. E. (1980). "Trajectory Divergence for Coupled Relaxation Oscillators: Measurements and Models," *J. Stat. Phys.* **23**(3), 321–333.

Golnaraghi, M., and Moon, F. C. (1985). "Chaotic Dynamics of a Nonlinear Servo Device," *24th IEEE Conf. on Decision and Control*, December 1985, Cornell University Report, Department of Theoretical and Applied Mechanics, July 1985.

Gorman, M., Widmann, P. J., and Robbins, K. A. (1984). "Chaotic Flow Regimes in a Convection Loop," *Phys. Rev. Lett.* **52**(25), 2241–2244.

Grassberger, P., and Proccacia, I. (1983). "Characterization of Strange Attractors," *Phys. Rev. Lett.* **50**, 346–349.

Grassberger, P. and Proccacia, I. (1984). "Dimensions and Entropies of Strange Attractors from a Fluctuating Dynamics Approach," *Physica* **13D**, 34–54.

Grebogi, C., Ott, E., and Yorke, J. A. (1983a). "Crises, Sudden Changes in Chaotic Attractors and Transient Chaos," *Physica* **7D**, 181–200.

Grebogi, C., Ott, E., and Yorke, J. A. (1983b). "Fractal Basin Boundaries, Long Lived Chaotic Transients and Unstable–Unstable Pair Bifurcation," *Phys. Rev. Lett.* **50**(13), 935–938.

Grebogi, C., Ott, E., Pelikan, S., and Yorke, J. A. (1984). "Strange Attractors that Are Not Chaotic," *Physica* **13D**, 261–268.

Grebogi, C., Ott, E., and Yorke, J. A. (1985a). "Attractors on an N-Torus: Quasiperiodicity Versus Chaos," *Physica* **15D**, 354–373.

Grebogi, C., Ott, E., and Yorke, J. A. (1985b). "Superpersistent Chaotic Transients," *Ergod. Theor. Dynam. Syst.* **5**, 341–372.

Grebogi, C., McDonald, S. W., Ott, E., and Yorke, J. A., (1985c). "Exterior Dimension of Fat Fractals," *Phys. Lett.* **110A**(1), 1–4.

Grebogi, C., Ott, E., and Yorke, J. A. (1986). "Metamorphoses of Basin Boundaries in Nonlinear Dynamical Systems," *Phys. Rev. Lett.* **56**(10), 1011–1014.

Guckenheimer, J. (1982). "Noise in Chaotic Systems," *Nature* **298**, 358.

Guckenheimer, J., and Holmes, P. J. (1983). *Nonlinear Oscillations, Dynamical Systems and Bifurcations of Vector Fields*, Springer-Verlag, New York.

Gwinn, E. G., and Westervelt, R. M. (1985). "Intermittent Chaos and Low-Frequency Noise in the Driven Damped Pendulum," *Phys. Rev. Lett.* **54**(15), 1613–1616.

Hao, B.-L. (1984). *Chaos*, World Scientific Publishers, Singapore.

Harrison, R. G. and Biswas, D. J. (1986). "Chaos in Light," *Nature* **321**, **22**, May, 394–401.

Hart, J. E. (1984). "A New Analysis of the Closed Loop Thermosyphon," *Int. J. Heat Mass Transfer* **27**(1), 125–136.

Haucke, H. and Ecke, R. E. (1987). "Mode-Locking and Chaos in Rayleigh-Benard Convection," *Physica D* (in press), Los Alamos National Lab. Report LA-UR-86-1257.

Hayashi, C. (1953). *Forced Oscillations in Nonlinear Systems*, Nippon Printing and Publishing Co., Osaka, Japan.

Hayashi, H., Ishizuka, S., Ohta, M., and Hirakawa, J. (1982). "Chaotic Behavior in the 'Onchidium' Giant Neuron under Sinusoidal Stimulation," *Phys. Lett.* **88A**(8), 5 April, 435–438.

Helleman, R. H. G. (1980a). "Self-Generated Chaotic Behavior in Nonlinear Mechanics," *Fund. Probl. Stat. Mech.* **5**, 165–233.

Helleman, R. H. G. (ed.) (1980b). "Nonlinear Dynamics," *Ann. N.Y. Acad. Sci.* **357**.

Hendriks, F. (1983). "Bounce and Chaotic Motion in Print Hammers," *IBM J. Res. Dev.* **27**(3), 273–280.

Henon, M. (1976). "A Two-Dimensional Map with a Strange Attractor," *Commun. Math. Phys.* **50**, 69.

Henon, M. (1982). "On the Numerical Computation of Poincaré Map," *Physica* **5D**, 412–414.

Henon, M., and Heiles, C. (1964). "The Applicability of the Third Integral of Motion: Some Numerical Experiments," *Astron. J.* **69**(1), 73–79.

Henon, M., and Wisdom, J. (1983). "The Benettin–Strelcyn Oval Billiard Revisited," *Physica* **8D**, 157–169.

Hockett, K., and Holmes, P. J. (1985). "Josephson Junction, Annulus Maps, Birkhoff Attractors, Horseshoes and Rotation Sets," Center for Applied Math Report Cornell University.

Holmes, P. J. (1979). "A Nonlinear Oscillator With a Strange Attractor," *Philos. Trans. R. Soc. London A* **292**, 419–448.

Holmes, P. J. (ed.) (1981). *New Approaches to Nonlinear Problems in Dynamics*, SIAM, Philadelphia, PA.

Holmes, P. J. (1982). "The Dynamics of Repeated Impacts With a Sinusoidally Vibrating Table," *J. Sound Vib.* **84**, 173–189.

Holmes, P. J. (1984b). "Bifurcation Sequences in Horseshoe Maps: Infinitely Many Routes to Chaos," *Phys. Lett. A* **104**(6, 7), 299–302.

Holmes, P. J. (1985). "Dynamics of a Nonlinear Oscillator with Feedback Control," *J. of Dynamics Systems, Measurements and Control* **107**, 159–165.

Holmes, P. J. (1986). "Chaotic Motions in a Weakly Nonlinear Model for Surface Waves," *J. Fluid Mech.* **162**, 365–388.

Holmes, P. J., and Moon, F. C. (1983). "Strange Attractors and Chaos in Nonlinear Mechanics," *J. Appl. Mech.* **50**, 1021–1032.

Holzfuss, J. and Mayer-Kress, G. (1986). "An Approach to Error-Estimation in the Application of Dimension Algorithms," *Dimensions and Entropies in Chaotic Systems*, G. Mayer-Kress (ed.), Springer-Verlag, Berlin.

Hudson, J. L., Mankin, J. C., and Rossler, O. E. (1984). "Chaos in Continuous Stirred Chemical Reaction," in *Stochastic Phenomena and Chaotic Behavior in Complex Systems*, P. Schuster (ed.), Springer-Verlag, Berlin, pp. 98–105.

Hunt, E. R., and Rollins, R. W. (1984). "Exactly Solvable Model of a Physical System Exhibiting Multidimensional Chaotic Behavior," *Phys. Rev. A* **29**(2), 1000–1002.

Hsu, C. S. (1981). "A Generalized Theory of Cell-to-Cell Mapping for Nonlinear Dynamical Systems," *J. Appl. Mech.* **48**, 634–642.

Hsu, C. S. (1987). *Cell to Cell Mapping*, Springer-Verlag.

Hsu, C. S., and Kim, M. C. (1985). "Statistics for Strange Attractors by Generalized Cell Mapping," *J. Stat. Phys.*

Iansiti, M., Hu, Q., Westervelt, R. M., and Tinkham, M. (1985). "Noise and Chaos in a Fractal Basin Boundary Regime of a Josephson Junction," *Phys. Rev. Lett.* **55**(7), August 12, 746–749.

Iooss, G., and Joseph, D. D. (1980). *Elementary Stability and Bifurcation Theory*, Springer-Verlag, New York.

Isomaki, H. M., von Boehm, J., and Raty, R. (1985). "Devil's Attractors and Chaos of a Driven Impact Oscillator," *Phys. Lett.* **107A**(8), 343–346.

Kadanoff, L. P. (1983). "Roads to Chaos," *Phys. Today* (Dec.), 46–53.

Kaplan, J., and Yorke, J. A. (1978). *Springer Lecture Notes in Mathematics*, No. 730, p. 228.

Keolian, R., Turkevich, L. A., Putterman, S. J., Rudnick, I., and Rudnick, J. A. (1981). "Subharmonic Sequences in the Faraday Experiment: Departures from Period Doubling," *Phys. Rev. Lett.* **47**(16), 1133–1511.

Klinker, T., Meyer-Ilse, X., and Lauterborn, W. (1984). "Period Doubling and Chaotic Behavior in a Driven Toda Oscillator," *Phys. Lett. A* **101**(8), 371–375.

Kobayashi, S. (1962). "Two-dimensional Panel Flutter 1. Simply Supported Panel," *Trans. Japan Society Aeronautical Space Sciences* **5**(8), 90–102.

Kostelich, E. J., and Yorke, J. A. (1985). "Lorenz Cross Sections and Dimension of the Double Rotor Attractor," *Dimensions and Entropies in Chaotic Systems*, G. Mayer-Kress (ed.), Springer-Verlag, Berlin, pp. 62–66.

Kostelich, E. J., Grebogi, C., Ott, E., and Yorke, J. A. (1987). "The Double Rotor Chaotic Attractor," to appear in *Physica D*.

Kreuzer, E. J. (1985). "Analysis of Chaotic Systems Using the Cell Mapping Approach," *Ingenieur-Archiv* **55**, 285–294.

Kuhn, T. (1962), *The Structure of Scientific Revolutions*, The University of Chicago Press, Chicago.

Landau, L. D. (1944). "On the Problem of Turbulence," *Akad. Nauk, Doklady* **44**, 339, Russian original reprinted in *Chaos* Hao, Bai-Lin (ed.) 107, World Scientific Publishers, Singapore.

Landauer, R. (1977). "Poor Man's Chaos," Internal Memo IBM Corp.

Lauterborn, W. (1981). "Subharmonic Route to Chaos in Acoustics," *Phys. Rev. Lett.* **47**(20), 1445–1448.

Lauterborn, W., and Cramer, E. (1981). "Subharmonic Route to Chaos Observed in Acoustics," *Phys. Rev. Lett.* **47**(20), 1145.

Lee, C.-K., and Moon, F. C. (1986). "An Optical Technique for Measuring Fractal Dimensions of Planar Poincaré Maps," *Phys. Lett. A* **114**(5), 222–226.

Leipnik, R. B. and Newton, T. A. (1981). "Double Strange Attractors in Rigid Body Motion with Linear Feedback Control," *Phys. Lett.* **86A**(2), 2, November, 63–67.

Levin, P. W., and Koch, B. P. (1981). "Chaotic Behavior of a Parametrically Excited Damped Pendulum," *Phys. Lett. A* **86**(2), 71–74.

Li, G. X. (1984). *Chaotic Vibrations of a Nonlinear System with Five Equilibrium States*, M.S. Thesis, Cornell University, Ithaca, N.Y., August.

Li, G. X. (1987). Doctoral dissertation, Department of Theoretical and Applied Mechanics, Cornell University, Ithaca, NY.

Libchaber, A. (1982). "Convections and Turbulence in Liquid Helium I," *Physica* **109 & 110B**, 1583–1589.

Libchaber, A. and Maurer (1978). *J. Phys. Lett.* **39**, 369.

Libchaber, A., Fauve, S., and Laroche, C. (1982). "Two-Parameter Study of the Routes to Chaos," in *Order in Chaos*, X. Campbell and X. Rose (eds.), North-Holland, Amsterdam.

Lichtenberg, A. J., and Lieberman, M. A. (1983). *Regular and Stochastic Motion*, Springer-Verlag, New York.

Lieberman, M. A., and Tsang, K. Y. (1985). "Transient Chaos in Dissipatively Perturbed Near-Integrable Hamiltonian Systems," *Phys. Rev. Lett.* **55**(9), 26, August, 908–911.

Lin, Y. K. (1976). *Probabilistic Theory of Structural Dynamics*, Krieger, Huntington, NY.

Linsay, P. A. (1981). "Period Doubling and Chaotic Behavior in a Driven, Anharmonic Oscillator," *Phys. Rev. Lett.* **47**(19), 1349–1352.

Linsay, P. S. (1985). The Structure of Chaotic Behavior in a PN Junction Oscillator, MIT, Department of Physics Report.

Lorenz, E. N. (1963). "Deterministic Non-Periodic Flow," *J. Atmos. Sci.* **20**, 130–141.

Lorenz, E. N. (1984). "The Local Structure of a Chaotic Attractor in Four Dimensions," *Physica* **13D**, 90–104.

Love, A. E. H. (1922). *A Treatise on the Mathematical Theory of Elasticity*, 4th Ed., Dover, New York.

L'vov, V. S., Predtechensky, A. A., and Chernykh, A. I. (1981). "Bifurcation and chaos in a system of Taylor vortices: a natural and numerical experiment," *Soviet Physics JETP* **53**, 562.

McDonald, S. W., Grebogi, C., Ott, E., and Yorke, J. A. (1985). "Fractal Basin Boundaries," *Physica* **17D**, 125–153.

McLaughlin, J. B. (1981). "Period-Doubling Bifurcations and Chaotic Motion for a Parametrically Forced Pendulum," *J. Stat. Phys.* **24**(2), 375–388.

Maganza, C. Caussé, and Laloë, F. (1986). "Bifurcations, Period Doubling and Chaos in Clarinetlike Systems," *Europhys. Lett.* **1**(6), 295–302.

Mahaffey, R. A. (1976). *Phys. Fluids* **19** 1387–1391.

Malraison, G., Atten, P., Bergé, P., and Dubois, M. (1983). "Dimension of Strange Attractors: An Experimental Determination of the Chaotic Regime of Two Convective Systems," *J. Phys. Lett.* **44**, 897–902.

Mandelbrot, B. B. (1977). *Fractals, Form, Chance, and Dimension*, W. H. Freeman, San Francisco, CA.

Manneville, P., and Pomeau, Y. (1980). "Different Ways to Turbulence in Dissipative Dynamical Systems," *Physica* **1D**, 219–226.

Marzec, C. J., and Spiegel, E. A. (1980). "Ordinary Differential Equations with Strange Attractors," *SIAM J. Appl. Math.* **38**(3), 403–421.

Matsumoto, T. (1984). "A Chaotic Attractor from Chua's Circuit," *IEEE. Trans. Circuits Syst.* **CAS-31**(12), 1055–1058.

Matsumoto, T., Chua, I. O., and Tanaka, S. (1984). "Simplest Chaotic Nonautonomous Circuit," *Phys. Rev. A.* **30**(2), 1155–1157.

Matsumoto, T., Chua, L. O., and Komuro, M. (1985). "The Double Scroll," *IEEE Trans. Circuits Syst.* **CAS-32**(8), 798–818.

May, R. M. (1976). "Simple Mathematical Models with Very Complicated Dynamics," *Nature* **261**, 459–467.

Mayer-Kress, G. (1985). "Introductory Remarks," in *Dimensions and Entropies in Chaotic Systems*, G. Mayer-Kress (ed.), Springer-Verlag, Berlin.

Miles, J. (1984a). "Resonant Motion of Spherical Pendulum," *Physica* **11D**, 309–323.

Miles, J. (1984b). "Resonantly Forced Motion of Two Quadratically Coupled Oscillators," *Physica* **13D**, 247–260.

Miles, J. (1984c). "Resonant, Nonplanar Motion of a Stretched String," *J. Acoust. Soc. Am.* **75**(5), 1505–1510.

Minorsky, N. (1962). *Nonlinear Oscillations*, Van Nostrand, Princeton, NJ.

Moon, F. C. (1980a). "Experiments on Chaotic Motions of a Forced Nonlinear Oscillator: Strange Attractors," *ASME J. Appl. Mech.* **47**, 638–644.

Moon, F. C. (1980b). "Experimental Models for Strange Attractor Vibration in Elastic Systems," in *New Approaches to Nonlinear Problems in Dynamics*, P. J. Holmes (ed.), pp. 487–495.

Moon, F. C. (1984a). *Magneto-Solid Mechanics*, Wiley, New York.

Moon, F. C. (1984b), "Fractal Boundary for Chaos in a Two State Mechanical Oscillator," *Phys. Rev. Lett.* **53**(60), 962–964.

Moon, F. C. (1986). "New Research Directions for Chaotic Phenomena in Solid Mechanics," in *Perspectives in Nonlinear Dynamics*, R. Cawley and M. Shlesinger (eds.), World Scientific Publishers, Singapore.

Moon, F. C., and Holmes, P. J. (1979). "A Magnetoelastic Strange Attractor," *J. Sound Vib.* **65**(2), 275–296; "A Magnetoelastic Strange Attractor," *J. Sound Vib.* **69**(2), 339.

Moon, F. C., and Holmes, W. T. (1985). "Double Poincaré Sections of a Quasi-Periodically Forced, Chaotic Attractor," *Phys. Lett. A* **111**(4), 157–160.

Moon, F. C., and Li, G.-X. (1985a). "The Fractal Dimension of the Two-Well Potential Strange Attractor," *Physica* **17D**, 99–108.

Moon, F. C., and Li, G.-X. (1985b). "Fractal Basin Boundaries and Homoclinic Orbits for Periodic Motion in a Two-Well Potential," *Phys. Rev. Lett.* **55**(14), 1439–1442.

Moon, F. C., and Shaw, S. W. (1983). "Chaotic Vibration of a Beam with Nonlinear Boundary Conditions," *J. Nonlinear Mech.* **18**, 465–477.

Moon, F. C., Cusumano, J., and Holmes, P. J. (1987). "Evidence for Homoclinic Orbits as a Precursor to Chaos in a Magnetic Pendulum," *Physica D* (in press).

Moore, D. W., and Spiegel, E. A. (1966). "A Thermally Excited Non-linear Oscillator," *Astrophys. J.* **143**(3), 871–887.

Nayfeh, A. H., and Mook, D. T. (1979). *Nonlinear Oscillations*, Wiley, New York.

Nayfeh, A. H. and Khdeir, A. A. (1986). "Nonlinear Rolling of Ships in Regular Beam Seas," *Int. Shipbuilding Prog.* **33**, No. 379, 40–49.

Newhouse, S., Ruelle, D., and Takens, F. (1978). "Occurrence of Strange Axiom A Attractors Near Quasiperiodic Flows on T^m, $m \geq 3$," *Commun. Math. Phys.* **64**, 35–40.

Ott, E. (1981). "Strange Attractors and Chaotic Motions of Dynamical Systems," *Rev. Mod. Phys.* **53**(4), Part 1, 655–671.

Packard, N. H., Crutchfield, J. P., Farmer, J. D., and Shaw, R. S. (1980). "Geometry from a Time Series," *Phys. Rev. Lett.* **45**, 712.

Peitgen, H. -O., and Richter, P. H. (1986). *The Beauty of Fractals*, Springer-Verlag, Berlin.

Pezeshki, C., and Dowell, E. H. (1987). "An Examination of Initial Condition Maps for the Sinusoidally Excited Buckled Beam Modeled by Duffing's Equation," *J. Sound and Vibration* (in press).

Poddar, B., Moon, F. C., and Mukherjee, S. (1986). "Chaotic Motion of an Elastic–Plastic Beam," *J. Appl. Mech.* (in press).

Poincaré, H. (1921). *The Foundation of Science: Science and Method*, English Translation, The Science Press, N.Y.

Pomeau, Y., and Manneville, P. (1980). "Intermittent Transition to Turbulence in Dissipative Dynamical Systems," *Commun. Math. Phys.* **74**, 189–197.

Prigogine, I. and Stengers, I. (1984). *Order Out of Chaos*, Bantam Books, Toronto.

Rand, D., Ostlund, S., Sethna, J., and Siggia, E. D. (1982). "Universal Transition From Quasiperiodicity to Chaos in Dissipative Systems," *Phys. Rev. Lett.* **49**(2), 387–390.

Richter, P. H., and Scholz, H.-J. (1984). "Chaos in Classical Mechanics: The Double Pendulum," in *Stochastic Phenomena and Chaotic Behavior in Complex Systems*, P. Schuster (ed.), Springer-Verlag, Berlin, pp. 86–97.

Robbins, K. A. (1977). "A New Approach to Subcritical Instability and Turbulent Transitions in a Simple Dynamo," *Math. Proc. Camb. Philos. Soc.* **82**, 309–325.

Rollins, R. W., and Hunt, E. R. (1982). "Exactly Solvable Model of a Physical System Exhibition Universal Chaotic Behavior," *Phys. Rev. Lett.* **49**(18), 1295–1298.

Rössler, O. E. (1976a). "Chemical Turbulence: Chaos in a Small Reaction-Diffusion System," *Z. Naturforsch. a* **31**, 1168–1172.

Rössler, O. E. (1976b). "An Equation for Continuous Chaos," *Phys. Lett. A* **57**, 397.

Roux, T. C., Simoyi, R. H., and Swinney, H. L. (1983). "Observation of a Strange Attractor," *Physica* **8D**, 257–266.

Russel, D. A., Hanson, J. D., and Ott, E. (1980). "Dimension of Strange Attractors," *Phys. Rev. Lett.* **45**(14), 1175–1178.

Saltzman, B. (1962). "Finite Amplitude Free Convection as an Initial Value Problem—I," *J. Atmos. Sci.* **19**, 329–341.

Schreiber, I., Kubicek, M., and Marak, M. (1980). "On Coupled Cells," in *New Approaches to Nonlinear Problems in Dynamics*, P. J. Holmes (ed.), SIAM, Philadelphia, PA, pp. 496–508.

Schuster, H. G. (1984). *Deterministic Chaos*, Physik-Verlag GmbH, Weinheim (F.R.G.).

Shaw, R. (1981). "Strange Attractors, Chaotic Behavior and Information Flow," *Z. Naturforsch. A* **36**, 80–112.

Shaw, R. (1984). *The Dripping Faucet as a Model Chaotic System*, Aerial Press, Santa Cruz, CA.

Shaw, S. W. (1985). "The Dynamics of a Harmonically Excited System Having Rigid Amplitude Constraints, Parts 1, 2," *J. Applied Mechanics* **52**(2), 453–464.

Shaw, S., and Holmes, P. J. (1983). "A Periodically Forced Piecewise Linear Oscillator," *J. Sound Vib.* **90**(1), 129–155.

Shimada, I., and Nagashima, T. (1979). "A Numerical Approach to Ergodic Problem of Dissipative Dynamical Systems," *Prog. of Theoretical Phys.* **61**(6), 1605–1616.

Simoyi, R. H., Wolf, A., and Swinney, H. L. (1982). "One-Dimensional Dynamics in a Multi-Component Chemical Reaction," *Phys. Rev. Lett.* **49**, 245.

Soong, T. T. (1973). *Random Differential Equations in Science and Engineering*, Academic Press, New York.

Sparrow, C. T. (1981). "Chaos in a Three-Dimensional Single Loop Feedback System With a Piecewise Linear Feedback Function," *J. Math. Anal. Appl.* **83**, 275–291.

Sparrow, C. (1982). *The Lorenz Equations: Bifurcations, Chaos, and Strange Attractors*, Springer-Verlag, New York.

Sreenivasan, K. R. (1986). "Chaos in Open Flow Systems," in *Dimension and Entropies*, in *Chaotic Systems*, G. Mayer-Kress (ed.), Springer-Verlag, New York.

Stoker, J. J. (1950). *Nonlinear Vibrations*, Interscience, New York.

Swinney, H. L. (1983). "Observations of Order and Chaos in Nonlinear Systems," in *Order and Chaos*, Campbell and Rose (eds.), North-Holland, Amsterdam, pp. 3–15.

Swinney, H. L. (1985). "Observations of Complex Dynamics and Chaos," in *Fundamental Problems in Statistical Mechanics VI*, E. G. D. Cohen (ed.), Elsevier Science Publishers, New York, pp. 253–289.

Swinney, H. L., and Gollub, J. P. (1978). "The Transition of Turbulence," *Physics Today*, **31**(8), 41 (August).

Symonds, P. S., and Yu, T. X. (1985). "Counterintuitive Behavior in a Problem or Elastic–Plastic Beam Dynamics," *J. Appl. Mech.* **52**, 517–522.

Szczygielski, W., and Schweitzer, G. (1985). "Dynamics of a High-Speed Rotor Touching a Boundary," $IUTAM/IFT_0MM$ *Symposium on the Dynamics of Multibody Systems*, CISM, Udine, Italy.

Szemplinska-Stupnicka, W., and Bajkowski, J. (1986). "The $\frac{1}{2}$ subharmonic Resonance and its Transition to Chaotic Motion in a Non-Linear Oscillator," *Int. J. Non-Linear Mechanics* (in press).

Tatsumi, T., (ed.) (1984). *Turbulence and Chaotic Phenomena in Fluids*, North-Holland, Amsterdam.

Termonia, Y., and Alexandrowicz, Z. (1983). *Phys. Rev. Lett.* **51**(14), 1265.

Testa, J., Perez, J., and Jeffries, C. (1982). "Evidence for Universal Chaotic Behavior in a Driven Nonlinear Oscillator," *Phys. Rev. Lett.* **48**, 714.

Thomson, W. T. (1965), *Vibration Theory*, Prentice Hall, Englewood Cliffs, New Jersey.

Thompson, J. M. T. (1983). "Complex Dynamics of Compliant Off-Shore Structures," *Proc. R. Soc. Land.* **A387**, 407–427.

Thompson, J. M. T., and Ghaffari, R. (1982). "Chaos After Period-Doubling Bifurcations in the Resonance of an Impact Oscillator," *Phys. Lett. A* **91**(1), 5–8.

Thompson, J. M. T., and Stewart, H. B. (1986). *Nonlinear Dynamics and Chaos*, Wiley, Chichester.

Tousi, S., and Bajaj, A. K. (1985). "Period-Doubling Bifurcations and Modulated Motions in Forced Mechanical Systems," *J. Appl. Mech.* **52**(2), 446–452.

Tseng, W.-Y., and Dugundji, J. (1971). "Nonlinear Vibrations of a Buckled Beam Under Harmonic Excitation," *J. Appl. Mech.* **38**, 467–476.

Tufillaro, N. B. and Albano, A. M. (1986). "Chaotic Dynamics of a Bouncing Ball," *Am. J. Phys.* **54**(10), 939–944.

Ueda, Y. (1979). "Randomly Transitional Phenomena in the System Governed by Duffing's Equation," *J. Stat. Phys.* **20**, 181–196.

Ueda, Y. (1980). "Steady Motions Exhibited By Duffing's Equation: a Picture Book of Regular and Chaotic Motions," *New Approaches to Nonlinear Problems in Dynamics*, P. J. Holmes (ed.) SIAM, Philadelphia, PA.

Ueda, Y., and Akamatsu, N. (1981). "Chaotically Transitional Phenomena in the Forced Negative Resistance Oscillator," *Proc. IEEE ISCA8 '80*. Also *IEEE Trans. Circuits Syst.* **CAS-28**(3), March.

Ueda, Y., Doumoto, H., and Nobumoto, K. (1978). "An Example of Random Oscillations in Three-Order Self-Restoring System," *Proceedings of the Electric and Electronic Communication Joint Meeting*, Kansai District, Japan, October.

Ueda, Y., Nakajima, H., Hikihara, T., and Stewart, H. B. (1986). *Forced Two-Well Potential Duffing Oscillator*.

Van Buskirk, R., and Jeffries, C. (1985a). "Observation of Chaotic Dynamics of Coupled Nonlinear Oscillators," *Phys. Rev. A* **31**(5), 3332–3357.

Van der Pol, B., and Van der Mark, J. (1927). "Frequency Demultiplication," *Nature* **120** (3019), 363–364.

Van Dyke, M. (1982). *An Album of Fluid Motion*, Parabolic Press, City.

Viet, O., Westfreid, X., and Guyon, E. (1983). "Art cinetique et chaos mëchanique," *Eur. J. Phys.* **4**, 74–76.

Virgin, L. N. (1986). "The Nonlinear Rolling Response of a Vessel Including Chaotic Motions Leading to Capsize in Regular Seas," *Applied Ocean Research*.

Wisdom, J., Peale, S. J. and Mignard, F. (1984). "The Chaotic Rotation of Hyperion," *Icarus*, **58**, 137–152.

Wolf, A. (1984). "Quantifying Chaos with Lyapunov Exponents," *Nonlinear Sci. Theory Appl.* Ed. A. V. Holden, Manchester Univ. Press.

Wolf, A., Swift, J. B., Swinney, H. L., and Vasano, J. A. (1985). "Determining Lyapunov Exponents from a Time Series," *Physica* **16D**, 285–317.

Wolfram, S. (1984). "Universality and Complexity in Cellular Automata," *Physica* **10D**, 1–35.

Wolfram, S. (1986). *Theory and Applications of Cellular Automata*, World Scientific Publ., Singapore.

Yorke, J. A., Yorke, E. D., and Maller-Paret, J. (1985). "Lorenz-like Chaos in Partial Differential Equations for a Heated Fluid Loop," University of Maryland Report.

Zaslavsky, G. M. (1978). "The Simplest Case of a Strange Attractor," *Phys. Lett. A* **69**(3), 145–147.

Zaslavsky, G. M., and Chirikov, B. V. (1972). "Stochastic Instability of Nonlinear Oscillations," *Sov. Phys. Usp.* **14**(5), 549–672.

Zhu, Z-X. (1983). "Experiment on the Chaotic Phenomena of an Upside Down Pendulum," Report of Laboratory of General Mechanics, Beijing University.

Author Index

Subject Index

305